T0132815

THE ILLUSTRATED FLORA OF ILLINOIS

The Illustrated Flora of Illinois

ROBERT H. MOHLENBROCK, General Editor

THE ILLUSTRATED FLORA OF ILLINOIS

FLOWERING PLANTS
basswoods to spurges

Robert H. Mohlenbrock

SOUTHERN ILLINOIS UNIVERSITY PRESS
Carbondale and Edwardsville

Library of Congress Cataloging in Publication Data
Mohlenbrock, Robert H., 1931–
  Flowering plants.

  (The Illustrated flora of Illinois)
  Bibliography: p.
  Includes index.
    1. Botany—Illinois. 2. Botany—Illinois—Pictorial
  works. 3. Dicotyledons—Identification. I. Title.
  II. Series: Illustrated flora of Illinois.
  QK157.M614   582.1309773   81–8585
  ISBN 0–8093–1025–2   AACR2

*Editorial expenses for this edition have been met in part by a grant
from the Joyce Foundation, administered by the Natural Land Institute, Rockford, Illinois.*

This book is dedicated to
Mark William Mohlenbrock,
my eldest son,
who has prepared all the illustrations
in this volume.

CONTENTS

ILLUSTRATIONS

FOREWORD

In 1967, the first volume of The Illustrated Flora of Illinois was published. That volume, which covered the ferns of Illinois, has been followed by eight volumes on flowering plants. This is the ninth book devoted to flowering plants, and the fourth one treating dicotyledonous plants. Several additional volumes on dicots will follow, as well as work on algae, mosses, liverworts, lichens, fungi, and the monocotyledonous genus *Carex*.

The idea of The Illustrated Flora, which was conceived in 1960, is to present every group of plants known to occur in Illinois. For each kind of plant, there will be a description, ecological notes, a distribution map, and illustrations showing the major features of the plant. Keys for easy identification of each kind of plant are presented.

An advisory board was created in 1964 to screen, criticize, and make suggestions for each volume of The Illustrated Flora during its preparation. The board is composed of botanists eminent in their area of specialty—Dr. Gerald W. Prescott, University of Montana (algae); Dr. Constantine J. Alexopoulos, University of Texas (fungi): Dr. Aaron J. Sharp, University of Tennessee (bryophytes); Dr. Rolla M. Tryon, Jr., The Gray Herbarium (ferns); and Dr. Robert F. Thorne, Rancho Santa Ana Botanical Garden (flowering plants).

The author is editor of the series and will prepare many of the volumes. Specialists in other groups are preparing volumes on plants of their special interest. As volumes are completed, they will be published since there is no special sequence for publication.

The author is proud to acknowledge the generous support of the Joyce Foundation which made possible the preparation of this volume. The Natural Land Institute of Rockford, Illinois, which sponsors The Illustrated Flora, provided administrative support.

Robert H. Mohlenbrock

Southern Illinois University
April 5, 1981

The Illustrated Flora of Illinois

FLOWERING PLANTS
basswoods to spurges

County Map of Illinois

Introduction

This volume is the fourth devoted to dicotyledons, or dicot plants. Dicots are the greatest group of flowering plants, exceeding the monocotyledons, or monocots. Dicots are plants which produce a pair of seed leaves during germination, while monocots produce merely a single seed leaf.

Five volumes have been published on monocots, covering such plants as grasses, sedges, lilies, orchids, irises, aroids, and pond-weeds.

The dicots include such plant groups as roses, mustards, mints, nightshades, milkweeds, asters, and pinks. The three previously published volumes on dicots treated hollies to loasas, willows to mustards, and magnolias to pitcher plants.

Since there are over 200,000 different kinds of flowering plants, it is necessary for systems of classifications to be developed to organize so many species. Since the time of Linnaeus in the eighteenth century, and even before, botanists have published hundreds of classification schemes. Each scheme has attempted to group plants of similar characters together. As more and more information is learned about plants, the newer systems of classification try to incorporate this information.

When I began to write the flowering plants in the series, I had to select one system to follow for the presentation of the flowering plants. After considerable soul searching, I chose a classification proposed by Robert Thorne in outline form in 1968. I have, however, departed from Thorne's system in a few instances. I am following Thorne in using the standard suffix -aceae for all families. Thus, the Cruciferae becomes the Brassicaceae, the Guttiferae becomes the Hypericaceae, the Leguminosae becomes the Fabaceae, the Umbelliferae becomes the Apiaceae, the Labiatae becomes the Lamiaceae, the Compositae becomes the Asteraceae, and the Gramineae becomes the Poaceae.

Since the Thorne classification is considerably different from the more traditional Engler system, the sequence for the dicots is presented next. Those names in boldface are described in this volume of The Illustrated Flora of Illinois.

Order Annonales
Family Magnoliaceae
Family Annonaceae
Family Calycanthaceae
Family Aristolochiaceae

Family Lauraceae
Family Saururaceae
Order Berberidales
Family Menispermaceae
Family Ranunculaceae

1

Family Berberidaceae
Family Papaveraceae
Order Nymphaeales
Family Nymphaeaceae
Family Ceratophyllaceae
Order Sarraceniales
Family Sarraceniaceae
Order Theales
Family Aquifoliaceae
Family Hypericaceae[1]
Family Elatinaceae
Family Ericaceae
Order Ebenales
Family Ebenaceae
Family Styracaceae
Family Sapotaceae
Order Primulales
Family Primulaceae
Order Cistales
Family Violaceae
Family Cistaceae
Family Passifloraceae
Family Cucurbitaceae
Family Loasaceae
Order Salicales
Family Salicaceae
Order Tamaricales
Family Tamaricaceae
Order Capparidales
Family Capparidaceae
Family Resedaceae
Family Brassicaceae
Order Malvales
Family Tiliaceae
Family Sterculiaceae
Family Malvaceae
Order Urticales
Family Ulmaceae
Family Moraceae
Family Urticaceae
Order Rhamnales
Family Rhamnaceae
Family Elaeagnaceae
Order Euphorbiales
Family Thymelaeaceae

Family Euphorbiaceae
Order Solanales
Family Solanaceae
Family Convolvulaceae
Family Polemoniaceae
Order Campanulales
Family Campanulaceae
Order Santalales
Family Celastraceae
Family Santalaceae
Family Loranthaceae
Order Oleales
Family Oleaceae
Order Geraniales
Family Linaceae
Family Zygophyllaceae
Family Oxalidaceae
Family Geraniaceae
Family Balsaminaceae
Family Limnanthaceae
Family Polygalaceae
Order Rutales
Family Rutaceae
Family Simaroubaceae
Family Anacardiaceae
Family Sapindaceae
Family Aceraceae
Family Hippocastanaceae
Family Juglandaceae
Order Myricales
Family Myricaceae
Order Chenopodiales
Family Phytolaccaceae
Family Nyctaginaceae
Family Aizoaceae
Family Cactaceae
Family Portulacaceae
Family Chenopodiaceae
Family Amaranthaceae
Family Caryophyllaceae
Family Polygonaceae
Order Hamamelidales
Family Hamamelidaceae
Family Platanaceae
Order Fagales
Family Fagaceae

[1]Called Clusiaceae by Thorne (1968).

Family Betulaceae
  Order Rosales
Family Rosaceae
Family Fabaceae
Family Crassulaceae
Family Saxifragaceae
Family Droseraceae
Family Staphyleaceae
  Order Myrtales
Family Lythraceae
Family Melastomaceae
Family Onagraceae
  Order Gentianales
Family Loganiaceae
Family Rubiaceae
Family Apocynaceae
Family Asclepiadaceae[2]
Family Gentianaceae
Family Menyanthaceae
  Order Bignoniales
Family Bignoniaceae
Family Martyniaceae
Family Scrophulariaceae
Family Plantaginaceae
Family Orobanchaceae

Family Lentibulariaceae
Family Acanthaceae
  Order Cornales
Family Vitaceae
Family Nyssaceae
Family Cornaceae
Family Haloragidaceae
Family Hippurdiaceae
Family Araliaceae
Family Apiaceae[3]
  Order Dipsacales
Family Caprifoliaceae
Family Adoxaceae
Family Valerianaceae
Family Dipsacaceae
  Order Lamiales
Family Hydrophyllaceae
Family Boraginaceae
Family Verbenaceae
Family Phrymataceae[4]
Family Callitrichaceae
Family Lamiaceae
  Order Asterales
Family Asteraceae

[2]Included in Apocynaceae by Thorne (1968).
[3]Included in Araliaceae by Thorne (1968).
[4]Included in Verbenaceae by Thorne (1968).

In this volume are four orders and ten families of dicots. Because such a small number of families is found in this work, no overall key to the dicot families of Illinois is included. For keys to all families of vascular plants in Illinois, my companion volume, *Guide to the Vascular Flora of Illinois* (1975), is recommended.

In this volume are the orders Malvales, Urticales, Rhamnales, and Euphorbiales. Within the Malvales are the families Tiliaceae, Sterculiaceae, and Malvaceae. The families Ulmaceae, Moraceae, and Urticaceae comprise the Urticales, while the Rhamnaceae and Elaeagnaceae make up the Rhamnales. The Euphorbiales includes only the Thymelaeaceae and the Euphorbiaceae.

The nomenclature for the species and lesser taxa used in this volume has been arrived at after lengthy study of recent floras and monographs. Synonyms, with complete author citation, which have applied to species in the northeastern United States, are given under each species. A description, while not necessarily intended to be complete, covers the more important features of the species.

The common name, or names, is the one used locally in Illinois. The habitat designation is not always the habitat throughout the range of the species, but only for it in Illinois. The overall range for each species is given from the northeastern to the northwestern extremities, south to the southwestern limit, then eastward to the southeastern limit. The range has been compiled from various sources, including examination of herbarium material and some field studies. A general statement is given concerning the range of each species in Illinois. Dot maps showing county distribution for each taxon are provided. Each dot represents a voucher specimen deposited in some herbarium. There has been no attempt to locate each dot with reference to the actual locality within each county.

The distribution has been compiled from field study as well as herbarium study. Herbaria from which specimens have been studied are the Field Museum of Natural History, Eastern Illinois University, the Gray Herbarium of Harvard University, Illinois Natural History Survey, Illinois State Museum, Missouri Botanical Garden, New York Botanical Garden, Southern Illinois University, the United States National Herbarium, the University of Illinois, and Western Illinois University. In addition, a few private collections have been examined. The author expresses his gratitude to the curators and staffs of these herbaria who helped him in his study.

Each species is illustrated, showing the habit as well as some of the distinguishing features in detail. Mark Mohlenbrock, my son, prepared all of the illustrations.

The Natural Land Institute and its director, Mr. George Fell, have provided the administrative support for this book which has been funded by the Joyce Foundation. I am also grateful to all the other people who have assisted me in this project. I wish to thank my daughter, Wendy Ann, for preparing all the maps. To my wife, Beverly, who has done all the clerical work from organizing my data to typing all drafts of the manuscript, I am deeply grateful.

Descriptions and Illustrations

Order Malvales

Of the six families which comprise the order Malvales, in accordance with the Thorne (1968) system of classification, three are represented in Illinois. The Tiliaceae and Malvaceae both have species native to the Illinois flora, while the Malvaceae and Sterculiaceae have species introduced into the Illinois flora. Not represented at all in Illinois are the Sphaerosepalaceae, Elaeocarpaceae, and Bombacaceae. This is essentially the same alignment of families that Cronquist (1968) recognizes, except that Cronquist transfers the Sphaerosepalaceae (= Rhopalocarpaceae) to his Theales, and adds the Scytopetalaceae to his Malvales.

The characters which hold the families of the Malvales together include a syncarpous pistil with a superior ovary and axile placentation, numerous centrifugal stamens connate by their filaments, free petals convolute in the bud, and a valvate calyx. In addition, except for the Elaeocarpaceae, most of the plants in the order have stellate pubescence or lepidote scales.

TILIACEAE–BASSWOOD FAMILY

Trees, shrubs, or infrequently herbs, mostly with branched hairs; leaves alternate, simple, stipulate; flowers perfect, actinomorphic, mostly in cymes; sepals usually 5, free or united at the base; petals 5, free, rarely absent; stamens 10–many, free or united at the base into fascicles of 5 or 10; pistil one, the ovary superior, 2- to 10-locular, with axile placentation, with 1–several ovules in each locule; fruits various.

This family contains about four hundred species. With the exception of *Tilia*, the other genera are very infrequently grown in cultivation.

Only the following genus occurs in Illinois.

1. *Tilia* L.–Basswood, Linden

Trees; leaves alternate, simple, serrate, the stipules deciduous; inflorescence cymose from a ligulate bract, the flowers perfect, actinomorphic; sepals 5, free; petals 5, free; stamens numerous; pistil one, the ovary superior, 5-locular, the style one, the stigma 5-toothed; fruit dry, globose, indehiscent, mostly 1-seeded.

The most recent monograph of the genus is by Jones (1968). The genus *Tilia* is sometimes grown as an ornamental. The most commonly grown ornamental is the native *T. americana*, although a few other species are planted in Illinois. The small-leaved basswood, *Tilia cordata* Mill., is a particularly attractive species.

KEY TO THE SPECIES OF Tilia IN ILLINOIS

1. Leaves glabrous beneath except in the axils of the veins _____
_____ 1. *T. americana*
1. Leaves pubescent beneath with stellate hairs _____ 2
    2. Leaves green beneath; peduncles and pedicels glabrous _____
_____ 1. *T. americana*
    2. Leaves white beneath; peduncles and pedicels pubescent _____
_____ 2. *T. heterophylla*

1. **Tilia americana** L. Sp. Pl. 514. 1753.

Tree rarely more than 30 m tall (in Illinois), up to 1 m in diameter, the crown round-topped; bark light brown, deeply furrowed at maturity; twigs gray to brown, slender, glabrous, with numerous dark lenticels; buds ovoid, dark red to light brown, glabrous, 4–6 mm long; leaves broadly ovate, acute to abruptly acuminate at the apex, cordate or truncate at the asymmetrical base, coarsely serrate, to 15 cm long, more than half as broad, the upper surface green, glabrous, lustrous, the lower surface yellow-green to gray, glabrous or stellate-pubescent throughout and with tufts of axillary hairs, the petioles glabrous, to 5 cm long; inflorescence cymose, pendulous, to 15-flowered, the peduncles glabrous, to 8 cm long, the bract obtuse to subacute at the apex, glabrous, to 10 cm long, to 3.5 cm broad; flowers 8–12 mm long, on essentially glabrous pedicels; sepals 5, free, ovate, glabrous to pubescent, 4–6 mm long; petals 5, free, lanceolate, white to pale yellow, 8–12 mm long; stamens numerous, arranged in five clusters with the filaments united at the base, with one oblong, petaloid staminodium opposite each petal and two-thirds as long; ovary villous; fruit ovoid to subglobose, 5–8 mm in diameter, pubescent.

Two varieties may be distinguished in Illinois.

1. Leaves glabrous beneath except in the axils of the veins _____
_____ 1a. *T. americana* var. *americana*

1. Leaves pubescent beneath with stellate hairs _____
   _____ 1b. *T. americana* var. *neglecta*

1a. Tilia americana L. var. **americana** *Fig. 1a–c.*

Tilia glabra Vent. Anal. Hist. Nat. Madrid 2:62. 1800.

Leaves glabrous beneath except in the axils of the veins.

COMMON NAMES: Basswood; Linden.

HABITAT: Rich woods and along streams; high dunes.

RANGE: Quebec to Manitoba and North Dakota, south to Texas and Alabama.

ILLINOIS DISTRIBUTION: Occasional to common throughout the state.

This is the common basswood in Illinois, differing from both var. *neglecta* and *T. heterophylla* by the absence of stellate pubescence on the lower surface of the leaves.

Basswood leaves are sometimes confused with larger, unlobed leaves of red mulberry. Red mulberry has different serration and pubescent patterns. In addition, red mulberry has some latex present and has three or more bud scales per bud.

The wood of the basswood is light but tough, making it ideal for boxes, toys, and certain kinds of furniture. The fibrous inner back is used in making ropes and mats.

In rural areas, the basswood is sometimes called bee tree because of the great number of honey bees attracted to the fragrant flowers.

Although basswood generally grows in rich woods and along streams, it also may be found on high dunes facing Lake Michigan.

Basswood is also prized as an ornamental.

The flowers bloom from May to July.

1b. Tilia americana L. var. **neglecta** (Spach) Fosberg, Castanea
   20:58. 1955. *Fig. 1d.*

Tilia neglecta Spach, Ann. Sci. Nat. II. 2:341, t. 15. 1834.

Leaves pubescent beneath with stellate hairs.

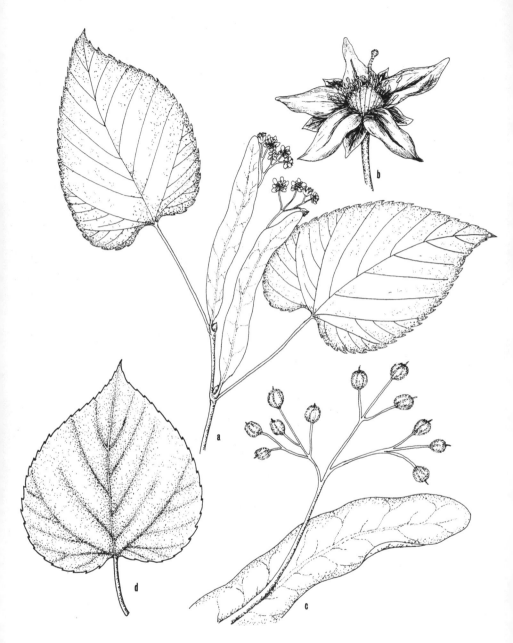

1. *Tilia americana* (Basswood). *a.* Flowering branch, × ½. *b.* Flower, × 2½. *c.* Bract, with fruits, × 1. var. *neglecta. d.* Leaf, × ½.

COMMON NAME: Basswood.

HABITAT: Rich woods.

RANGE: Quebec to Minnesota, south to Oklahoma and North Carolina.

ILLINOIS DISTRIBUTION: Rare; apparently confined to a few northern counties. This variety is sometimes considered to be a distinct species. However, since the only significant difference is in the presence of stellate pubescence on the lower surface of the leaves, I prefer to treat this taxon as a variety of *T. americana*.

2. **Tilia heterophylla** Vent. Anal. Hist. Nat. Madrid 2:63. 1800. *Fig. 2.*

Tilia michauxii Nutt. Sylva 1:92. 1845.

Tilia heterophylla Vent. var. *michauxii* (Nutt.) Sarg. Bot. Gaz. 66:506. 1818.

Tree up to 20 m tall (in Illinois), up to 0.65 m in diameter, the crown round-topped; bark light brown, furrowed at maturity; twigs yellow-brown to reddish, slender, glabrous; buds ovoid, reddish, glabrous except for the ciliate outer scales, to 6 mm long; leaves ovate, acute to abruptly acuminate at the apex, truncate or subcordate at the asymmetrical base, sharply serrate, to 12 cm long, at least half as broad, the upper surface dark green and glabrous at maturity, the lower surface covered with a dense white mat of stellate hairs, the petioles glabrous, to 4 cm long; inflorescence cymose, pendulous, 10- to 20-flowered, the peduncles pubescent, the bract obtuse at the apex, cuneate and asymmetrical at the base, becoming glabrous at maturity, to 15 cm long, to 3 cm broad; sepals 5, free, lance-ovate, pubescent, 4–6 mm long; petals 5, free, lanceolate, white to pale yellow, 6–8 mm long; stamens numerous, arranged in five clusters with the filaments united at the base, with one ovoid, petaloid staminodium opposite each petal and about two-thirds as long; ovary villous; fruit ellipsoid, 5–8 mm long, rusty-brown tomentose.

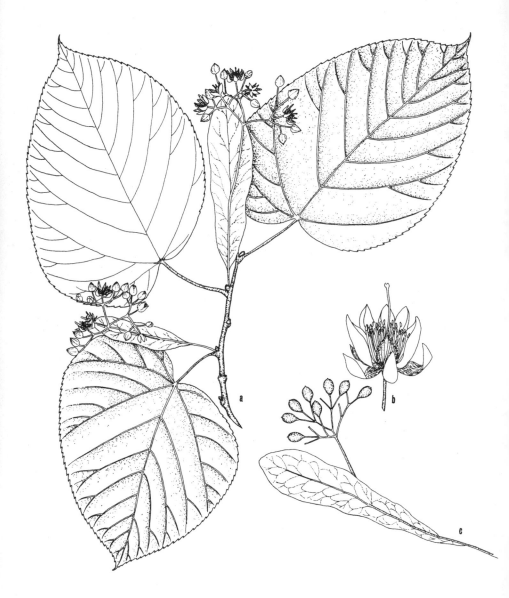

2. *Tilia heterophylla* (White Basswood). *a*. Flowering branch, × ½. *b*. Flower, with petals spread, × 2½. *c*. Bract, with fruits, × ½.

COMMON NAME: White Basswood.

HABITAT: Rich woods.

RANGE: New York across Ohio to Missouri, south to Arkansas and Florida.

ILLINOIS DISTRIBUTION: Known only from Hardin, Massac, and Pope counties.

The white basswood is a southeastern species which just reaches the southern tip of Illinois. It is distinguished from *T. glabra* var. *neglecta* by the dense mat of white stellate hairs on the lower surface of the leaves.

The report of *T. heterophylla* (as *T. floridana*) from White County by Palmer (1921) could not be verified.

The flowers bloom from May to July.

STERCULIACEAE–CHOCOLATE FAMILY

Trees, shrubs, or herbs with stellate pubescence; leaves alternate, simple or compound, the stipules caducous; flowers mostly perfect and actinomorphic; sepals 3–5, united at the base; petals none or very small; stamens usually 10, in 2 whorls, the filaments free or united, the outer five usually reduced to staminodia; pistil one, the ovary superior, 4- to 5-locular, with axile placentation, with 2 or more ovules per locule; fruits various.

This is primarily a tropical family represented by about 750 species. There are less than a dozen species native to the United States, in Florida, Arizona, and California. Important members of the family include *Theobroma cacao* L., the chocolate plant, and *Cola acuminata* Schott & Endl., the cola nut, as well as ornamentals such as *Brachychiton*, the bottle-trees, *Firmiana*, the Chinese parasol tree, and others.

Only the following genus has been found in Illinois.

1. Melochia L.

Herbs or shrubs, usually with stellate hairs; leaves alternate, simple, toothed; flowers small, variously arranged; calyx 5-lobed; petals 5, free; stamens 5, opposite the petals, united at the base; ovary superior, 5-locular; styles 5; fruit a 5-valved capsule.

There are approximately sixty species, most of them native to the Americas. Only the following species has been found as an adventive in Illinois.

3. *Melochia corchorifolia* (Chocolate Weed). *a.* Habit, × ½. *b.* Flower, × 5. *c.* Bract, × 10. *d.* Capsule, with bracts, × 5. *e.* Segment of capsule, × 5. *f.* Seed, × 5.

1. **Melochia corchorifolia** L. Sp. Pl. 675. 1753. *Fig.* 3.

Annual from a slender taproot; stems erect, up to 1 m tall, sparsely stellate-pubescent; leaves ovate to ovate-lanceolate, occasionally obscurely 3-lobed, acute at the apex, truncate to rounded to sub-cordate at the base, doubly serrate, glabrous except for some pubescence on the veins beneath, to 7 cm long, to 5 cm wide; petioles to 5 cm long, stellate-pubescent; stipules linear to linear-lanceolate, to 6 mm long; inflorescence composed of crowded, headlike cymes, with linear to linear-lanceolate bracts; sepals 5, green, united below, 2–4 mm long; petals 5, purple, free, 5–7 mm long; capsule 5-parted, subglobose, up to 5 mm in diameter, with 1 seed per locule; seed brown, wrinkled, 2.0–2.5 mm long.

COMMON NAME: Chocolate Weed.

HABITAT: Disturbed area along highway (in Illinois.)

RANGE: North Carolina to Mississippi, south into the tropics; adventive in Illinois.

ILLINOIS DISTRIBUTION: Known from Randolph County (along Illinois Route 3, south of Chester, near the bridge over Mary's River, *R. A. Evers*).

This annual species of a mostly tropical family has been found as an adventive one time in Illinois. This species is fairly common in cultivated fields of the southeastern United States. The flowers bloom in August and September.

MALVACEAE–MALLOW FAMILY

Herbs or shrubs, rarely trees; leaves alternate, simple, usually palmately veined, with small deciduous stipules; flowers usually perfect, rarely dioecious, actinomorphic; sepals mostly 5, usually united at least at the base; petals 5, free; stamens numerous, united into a central column around the pistil and attached to the base of the petals, the anthers 1-locular; pistil one, the ovary superior, several-locular, the ovules one to several per locule, the styles united below, free above and protruding beyond the staminal column; fruit a capsule or berry, several-locular.

The Malvaceae include nearly fifty genera and about one thousand species in most tropical and temperate regions of the World. They are easily recognized by the staminal column which surrounds the styles.

Of the eleven genera and twenty-four species in Illinois, only five genera and eight species are native.

KEY TO THE GENERA OF Malvaceae IN ILLINOIS

1. Some or all the leaves lobed _____ 2
1. None of the leaves lobed _____ 13
  2. Calyx subtended by an involucre of bracts at the base _____ 3
  2. Calyx without an involucre of bracts at the base _____ 11
3. Bracts at base of calyx 3 _____ 4
3. Bracts at base of calyx 6 or more _____ 10
  4. Bracts laciniate _____ 4. Gossypium
  4. Bracts unlobed, or at least not laciniate _____ 5
5. All or most of the leaves divided nearly to the base _____ 6
5. None of the leaves divided nearly to the base _____ 7
  6. Plants hispid; stems more or less procumbent _____ 2. Callirhoë
  6. Plants with scattered pubescence, but not hispid; stems ascending _____ 1. Malva
7. Flowers over 2 cm across _____ 8
7. Flowers less than 2 cm across _____ 1. Malva
  8. Lower leaves triangular, unlobed _____ 2. Callirhoë
  8. Lower leaves shallowly lobed, or, if unlobed, not triangular ___ 9
9. Bracts at base of calyx lanceolate to ovate-lanceolate; calyx lobes acute; carpels 1-seeded _____ 1. Malva
9. Bracts at base of calyx linear; calyx lobes acuminate; carpels 2- to 4-seeded _____ 3. Iliamna
  10. Bracts triangular; fruit separating at maturity into 15–20 carpels _____ 7. Althaea
  10. Bracts linear; fruit a 5-celled capsule _____ 6. Hibiscus
11. Flowers unisexual, white, up to 2 cm across _____ 8. Napaea
11. Flowers bisexual, blue or, if white, over 2 cm across _____ 12
  12. Flowers over 4 cm across _____ 2. Callirhoë
  12. Flowers up to 4 cm across _____ 9. Anoda
13. Leaves as broad as long, some of them at least 3 cm broad _____ 14
13. Leaves longer than broad, never 3 cm broad _____ 17
  14. Calyx not subtended by bracts _____ 10. Abutilon
  14. Calyx subtended by bracts _____ 15
15. Bracts subtending the calyx 3 _____ 16
15. Bracts subtending the calyx 6 or more _____ 6. Hibiscus
  16. Flowers over 2 cm broad, deep purple _____ 2. Callirhoë
  16. Flowers much less than 2 cm broad, pale lilac or white 1. Malva
17. Calyx subtended by 2–3 setaceous bracts _____ 5. Sphaeralcea
17. Calyx not subtended by bracts _____ 11. Sida

1. *Malva* L.–Mallow

Herbs; leaves alternate, simple, lobed or dissected; flowers axillary or terminal, solitary or clustered, perfect, actinomorphic, subtended by an involucel of 0–3 bracts; sepals 5, united; petals 5, free; stamen column anther-bearing only at the tip; pistil several-carpellate, each carpel 1-ovulate; fruiting carpels arranged in a circle, indehiscent, 1-seeded, beakless.

KEY TO THE TAXA OF Malva IN ILLINOIS

1. Flowers at least 2.5 cm across, usually broader; petals more than twice as long as the calyx _____ 2
1. Flowers up to 1.5 cm across; petals at most only twice as long as the calyx _____ 3
  2. Petals up to four times as long as the calyx; leaves shallowly lobed, the tips of the lobes triangular or broadly rounded; carpels about 10, minutely pubescent to nearly glabrous _____ 1. *M. sylvestris*
  2. Petals 5–8 times as long as the calyx; leaves deeply divided into narrow divisions, the tips slender and not triangular; carpels usually 15–20, densely pubescent _____ 2. *M. moschata*
3. Petals about twice as long as the calyx; leaves with rounded lobes or sometimes without lobes; carpels rounded on the margins _____ 4
3. Petals about as long as the calyx; leaves with angular lobes; carpels with sharp margins _____ 5. *M. rotundifolia*
  4. Leaves scarcely lobed, flat on the margins; stems procumbent _____ 3. *M. neglecta*
  4. Leaves with 5–7 rounded lobes, crisped on the margins; stems erect _____ 4. *M. verticillata* var. *crispa*

1. Malva sylvestris L. Sp. Pl. 689. 1753.

Biennial herb from thickened roots; stems erect to ascending, branched, to 75 cm tall, hirsute to nearly glabrous; leaves orbicular to reniform, to 10 cm across, truncate to subcordate at the base, crenate, 5- to 7-lobed, the lobes broadly rounded to triangular, hirsute to nearly glabrous, the petioles to 15 cm long; flowers clustered from the upper axils, to 5 cm across, 3-bracteate, the bracts oblong to lance-ovate, 5–12 mm long; sepals lance-ovate, green, glabrous or nearly so, 1–2 cm long; petals obcordate, reddish-purple with darker veins, 2–4 cm long; carpels mostly 10, arranged in a circle, flat on the back, rugose-reticulate, glabrous or minutely pubescent. Two varieties occur in Illinois.

1. Lobes of leaf triangular; stems and leaves hirsute _____
_____ 1a. *M. sylvestris* var. *sylvestris*
1. Lobes of leaf broadly rounded; stems and leaves glabrous or nearly
so _____ 1b. *M. sylvestris* var. *mauritiana*

1a. Malva sylvestris L. var. sylvestris *Fig. 4a*.

Lobes of leaf triangular; stems and leaves hirsute.

COMMON NAME: High Mallow.
HABITAT: Waste places.
RANGE: Native of Europe; escaped from cultivation from Quebec to North Dakota, south to Texas and Florida.
ILLINOIS DISTRIBUTION: Scattered in the northern four-fifths of the state.
The typical variety of *Malva sylvestris* has pointed leaf lobes and hirsute stems and leaves.
The large, showy flowers have made this plant a showy garden favorite.
The flowers bloom in August and September.

1b. Malva sylvestris L. var. mauritiana (L.) Boiss. Fl. Or. 1:819. 1867. *Fig. 4b–e*.

Malva mauritiana L. Sp. Pl. 689. 1753.

Lobes of leaf broadly rounded; stems and leaves glabrous or nearly so.

COMMON NAME: High Mallow.
HABITAT: Waste ground.
RANGE: Native of Europe; escaped from cultivation from New England to North Dakota, south to Texas and Florida.
ILLINOIS DISTRIBUTION: Known only from Cook County.
This variety, thought originally by Linnaeus to represent a distinct species, is less common in Illinois than var. *sylvestris*. It is occasionally grown as a garden ornamental, but rarely escapes from cultivation.
It blooms during August and September.

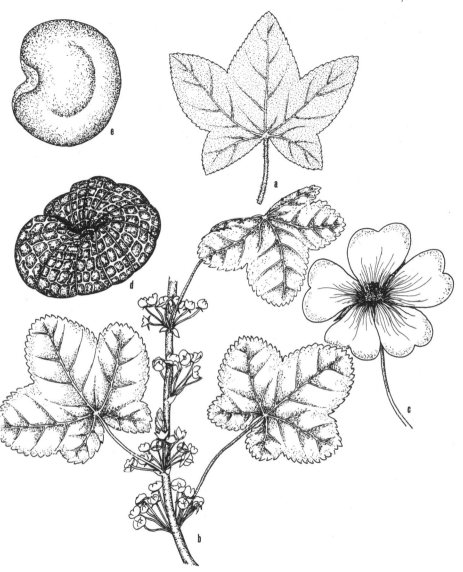

4. Malva sylvestris (High Mallow). *a*. Leaf, ×½. var. *mauritiana*. *b*. Leafy branch, with flowers, ×½. *c*. Flower, ×1. *d*. Fruiting carpel, ×5. *e*. Seed, ×10.

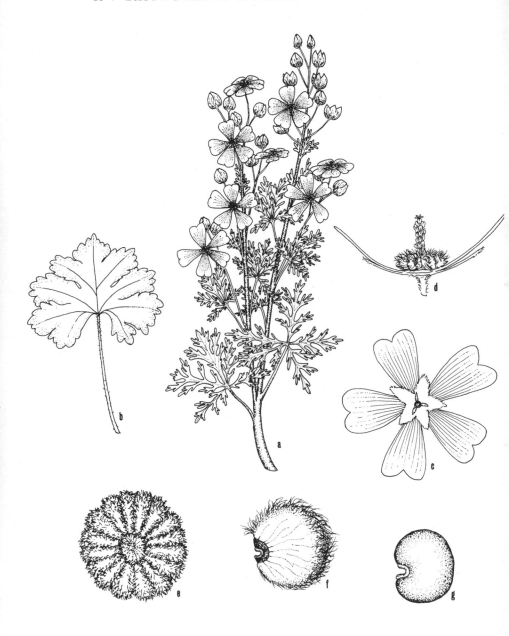

5. *Malva moschata* (Musk Mallow). *a.* Upper part of plant, ×½. *b.* Leaf variation, ×1. *c.* Flower, from below, ×1½. *d.* Carpels and staminal column, ×1½. *e.* Fruiting carpels, ×3. *f.* Fruiting carpel, ×5. *g.* Seed, ×5.

2. **Malva moschata** L. Sp. Pl. 690. 1753. *Fig. 5.*

Malva moschata L. var. *alba* Huett, Fl. LaSallensis 1:55. 1897.

Malva moschata L. f. *alba* (Huett) Moldenke, Phytologia 2:138. 1946.

Perennial herb; stems erect to ascending, branched, to 60 cm tall, pubescent; basal leaves and lower cauline leaves orbicular, to 10 cm across, shallowly 5- to 9-lobed, crenate, sparsely pubescent or nearly glabrous, or sometimes cleft like the upper cauline leaves, the petiole to 10 cm long; upper cauline leaves deeply palmately divided, the ultimate segments linear and pinnatifid, slightly pubescent to glabrous; flowers in terminal racemes, musk-scented, 2.5–3.5 cm across, 3-bracteate, the bracts oblanceolate, 4–10 mm long; sepals lance-ovate, green, glabrous or nearly so, 5–8 mm long; petals obcordate, pink or white, 2–3 cm long; carpels 15–20, arranged in a circle, rounded on the back, densely pubescent.

COMMON NAME: Musk Mallow.

HABITAT: Along railroads; in fields.

RANGE: Native of Europe; naturalized from Newfoundland to Ontario, south to Nebraska, Tennessee, and Delaware.

ILLINOIS DISTRIBUTION: Not common; known only from Cook, DuPage, LaSalle, and Lawrence counties. The common name for this species is derived from the faint musky odor of the flowers which bloom from June to August.

The LaSalle County record is based on a specimen with white flowers.

The rather large flowers and the deeply divided leaves distinguish this mallow from all others in Illinois.

3. **Malva neglecta** Wallr. Syll. Pl. Nov. Ratisb. 1:140. 1824. *Fig. 6.*

Biennial herb from a deep root; stems procumbent and spreading, branched, to 30 cm long, glabrous or nearly so; leaves orbicular to reniform, to 6 cm across, crenate or very shallowly lobed, glabrous or nearly so, the petioles to 12 cm long; flowers clustered in the axils of the upper leaves, to 1.5 cm across, 3-bracteate, the bracts lanceolate, 2–4 mm long; sepals lanceolate, green, glabrous, 4–7 mm long; petals obcordate, 8–12 mm long, white to pale lilac; carpels about 15, arranged in a circle, rounded on the back, puberulent.

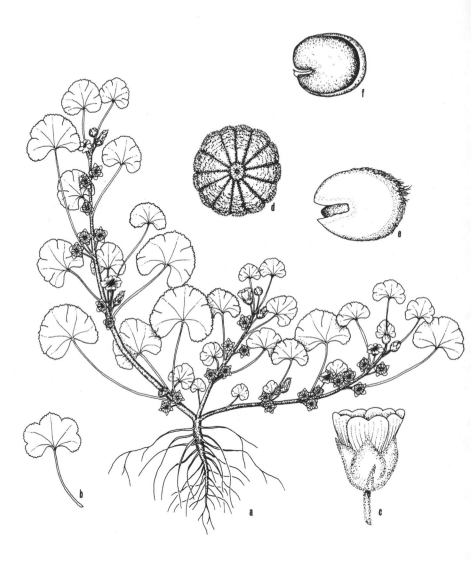

6. *Malva neglecta* (Common Mallow). *a*. Habit, × ½. *b*. Leaf variation, × ½. *c*. Flower, × 2½. *d*. Fruiting carpels, × 5. *e*. Fruiting carpel, × 7½. *f*. Seed, × 7½.

COMMON NAME: Common Mallow.

HABITAT: Barnyards, fields, lawns, along roads, along railroads.

RANGE: Native of Europe; naturalized throughout North America.

ILLINOIS DISTRIBUTION: Occasional to common throughout the state.

This is the common species of *Malva* in Illinois. For years, it was erroneously known as *M. rotundifolia*, but this binomial actually applies to a different species in Illinois.

Malva neglecta differs from *M. rotundifolia* by its petals twice as long as the calyx and the carpels which are rounded on the back. It differs from *M. verticillata* var. *crispa* by having flat leaf margins, obscurely lobed leaves, and procumbent stems.

The flowers bloom from June to September.

4. **Malva verticillata** L. var. **crispa** L. Sp. Pl. 689. 1753. *Fig. 7.*

Malva crispa (L.) L. Sp. Pl., ed. 2, 970. 1763.

Annual herb from fibrous roots; stems erect, branched, to 2 m tall, glabrous or nearly so; leaves orbicular, to 8 cm across, glabrous or nearly so, shallowly 5- to 11-lobed, the lobes crenate, crisped along the margin, the petioles to 10 cm long; flowers clustered in the axils of the upper leaves, to 1.5 cm across, 3-bracteate, the bracts lanceolate, 2–4 mm long; sepals lanceolate, green, glabrous, 4–7 mm long; petals obcordate, 8–12 mm long, white; carpels about 15, arranged in a circle, rugose-reticulate, glabrous or nearly so.

COMMON NAME: Curled Mallow.

HABITAT: Waste ground.

RANGE: Native of Europe; adventive in northeastern North America.

ILLINOIS DISTRIBUTION: Rarely escaped from cultivation in a few of the northern counties.

This mallow is distinctive by the crisped margins of the leaves. It differs further from *M. neglecta* by its erect stature.

The flowers bloom from July to September.

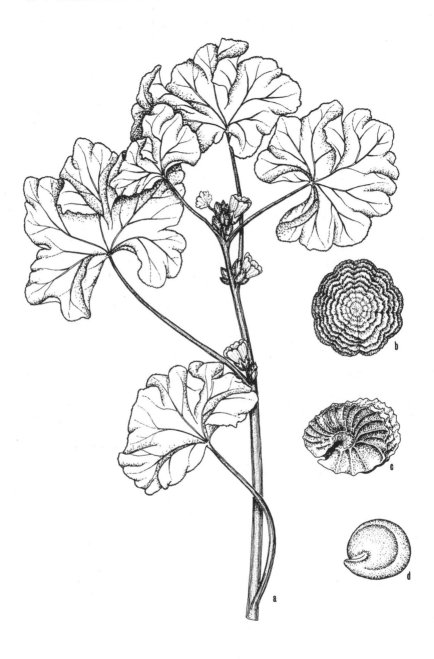

7. *Malva verticillata* var. *crispa* (Curled Mallow). *a*. Branch, with leaves and flowers, ×½. *b*. Fruiting carpels, ×5. *c*. Fruiting carpel, ×7½. *d*. Seed, ×7½.

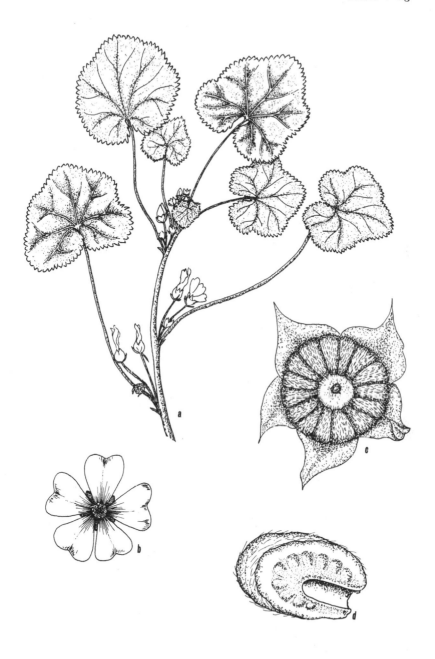

8. *Malva rotundifolia* (Mallow). *a*. Leafy branch, with flowers, × ½. *b*. Flower, × 2. *c*. Fruiting carpels, subtended by sepals, × 5. *d*. Fruiting carpels, × 10.

5. **Malva rotundifolia** L. Sp. Pl. 688. 1753. *Fig. 8.*

Biennial herb from a deep root; stems erect to ascending, branched, to 30 cm tall, glabrous or nearly so; leaves orbicular to reniform, to 6 cm across, shallowly lobed, crenate, glabrous or nearly so, the petioles to 12 cm long; flowers clustered in the axils of the upper leaves, to 1.5 cm across, 3-bracteate, the bracts lanceolate, 2–4 mm long; sepals lanceolate, green, glabrous, 6–10 mm long; petals obcordate, 7–12 mm long, white to pale lilac, the claws pubescent; carpels about 15, arranged in a circle, sharp-margined, pubescent at first, becoming glabrous or nearly so.

COMMON NAME: Mallow.

HABITAT: Barnyards.

RANGE: Native of Europe; sparsely naturalized in the western half of the United States.

ILLINOIS DISTRIBUTION: Scattered in the state, but very uncommon.

This species, often confused with *M. neglecta*, differs by its petals and sepals being subequal and by the sharp margins of the fruiting carpels.

The flowers bloom from June to September.

2. *Callirhoë* Nutt.—Poppy Mallow

Herbs; leaves alternate, unlobed to lobed or deeply parted; flowers axillary or terminal, solitary or clustered, perfect, actinomorphic, with or without an involucel of bracts; sepals 5, united at the base; petals 5, free; stamen column anther-bearing only at the tip; pistil several-carpellate, each carpel 1-ovulate; fruiting carpels arranged in a circle, indehiscent or 2-valved, 1-seeded, beaked.

Callirhoë is primarily a western North American genus of nine usually beautiful herbs.

KEY TO THE SPECIES OF Callirhoë IN ILLINOIS

1. Each flower subtended by three bracts _____ 2
1. Flowers without subtending bracts _____ 3
   2. Some of the leaves triangular; flowers in panicles or appearing umbellate; carpels not rugose _____ 1. *C. triangulata*
   2. None of the leaves triangular; flower solitary; carpels rugose _____
   _____ 2. *C. involucrata*
3. Basal leaves more or less triangular; top of carpels pubescent _____
   _____ 3. *C. alcaeoides*
3. Basal leaves rounded; top of carpels glabrous _____ 4. *C. digitata*

9. *Callirhoë triangulata* (Poppy Mallow). *a.* Leafy branch, with flower, × ½. *b.* Leaf, × 1. *c.* Flower, × 1. *d.* Fruiting carpels, × 5. *e,f.* Fruiting carpel, × 5.

1. **Callirhoë triangulata** (Leavenw.) Gray, Mem. Am. Acad. II.
4:16. 1848. *Fig. 9*.

Malva triangulata Leavenw. Am. Journ. Sci. 7:62. 1824.
Malva houghtonii Torr. & Gray, Fl. N. Am. 1:225. 1838.

Perennial herb from a deep root; stems erect or ascending, branched, to 60 cm tall, stellate-pubescent; leaves triangular, obtuse to acute at the apex, truncate to subcordate at the base, crenate, sometimes hastate, to 8 cm long, stellate-pubescent, the uppermost leaves often 3- to 5-parted, on shorter petioles; flowers in terminal panicles, 2.2–4.5 cm across, subtended by three involucral bracts, the bracts linear to spatulate, 3–6 mm long; sepals 5, united at the base, 4–7 mm long, stellate-pubescent; petals 5, free, obovate, purple, 1.7–3.2 cm long; fruiting carpels numerous, short-beaked, pubescent, not rugose.

COMMON NAME: Poppy Mallow.
HABITAT: Sandy soil, often in prairies.
RANGE: North Carolina to Wisconsin and Nebraska, south to Texas and Alabama.
ILLINOIS DISTRIBUTION: Confined to the northern half of the state. There is also one record each from Lawrence and Macoupin counties.
This beautiful species is distinguished from the similar *C. involucrata* by its sometimes triangular leaves and from *C. alcaeoides* by its three bracts at the base of each flower. Almost all parts of the plant are densely stellate-pubescent.

There is considerable variation in the shape of the leaves. The uppermost leaves tend to be divided into 3–5 lobes, while the lower leaves are usually triangular and undivided.

The poppy mallow is invariably found in sandy soils of prairies and rocky woods. It also is found in sandy cemeteries.

Torrey and Gray's *Malva houghtonii*, recorded from Illinois by Engelmann (1843) and Mead (1846), is the same species.

The flowers bloom from June to September.

2. **Callirhoë involucrata** (Torr. & Gray) Gray, Mem. Am. Acad. II. 4:16. 1848. *Fig. 10*.

Malva involucrata Torr. & Gray, Fl. N. Am. 1:226. 1838.

Perennial herb from a deep root; stems procumbent but turned upward at the tips, branched, to 60 cm long, hispid: leaves orbicular in outline, palmately 5- to 7-lobed, cordate at the base, the lobes

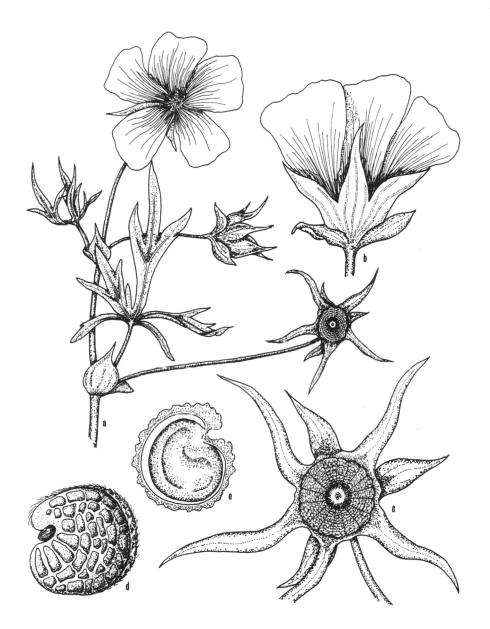

10. *Callirhoë involucrata* (Poppy Mallow). *a*. Leafy branch, with flower and fruit, ×¾. *b*. Flower, ×1.½ *c*. Fruit, ×1½. *d*. Fruiting carpel, ×10. *e*. Fruiting carpel, split lengthwise, showing seed, ×10.

linear to lanceolate, dentate or entire, hispid, the petioles hispid; stipules persistent, ovate; flowers solitary from the upper axils, 2.0–4.5 cm broad, on hispid pedicels to 10 cm long, subtended by three involucral bracts, the bracts linear, 5–8 mm long; sepals 5, united at the base, the lobes lanceolate, green, pubescent, 1.0–1.5 cm long; petals 5, free, spatulate, reddish-purple, 2–3 cm long; fruiting carpels numerous, short-beaked, pubescent, rugose-reticulate.

COMMON NAME: Poppy Mallow.

HABITAT: Along railroads and roads; in fields.

RANGE: Native to the western United States; adventive east as far as Ohio.

ILLINOIS DISTRIBUTION: Adventive in a few counties in the northern half of the state.

This western species is a showy adventive in fields and along railroads in a few northern counties. It can be distinguished from the similar-appearing *C. triangulata* by having all its leaves divided.

In Illinois, this species blooms from June to August.

3. **Callirhoë alcaeoides** (Michx.) Gray, Mem. Am. Acad. II. 4:18. 1848. *Fig. 11*.

Sida alcaeoides Michx. Fl. Bor. Am. 2:44. 1803.

Perennial herb from a thick, woody root; stems erect, slender, several-branched from the base, to 45 cm tall, strigose; basal leaves triangular, obtuse at the apex, cordate at the base, palmately lobed or divided, crenate, strigose, to 8 cm long, the petioles strigose; upper leaves palmately divided, often nearly to the base, the segments linear; flowers in terminal corymbs, 2.0–2.5 cm broad, on strigose pedicels to 5 cm long, without involucral bracts; sepals 5, united at the base, the lobes triangular, 3–4 mm long; petals 5, free, cuneate, pink or white, 5–10 mm long; fruiting carpels numerous, short-beaked, strigose, rugose-reticulate.

COMMON NAME: Pale Poppy Mallow.

HABITAT: Dry, gravelly areas.

RANGE: Illinois to Nebraska, south to Texas and Alabama.

ILLINOIS DISTRIBUTION: Known only from Cass, Christian, Henry, Peoria, and Winnebago counties.

The pale poppy mallow is readily recognized because of its smaller and paler flowers. The strigose pubes-

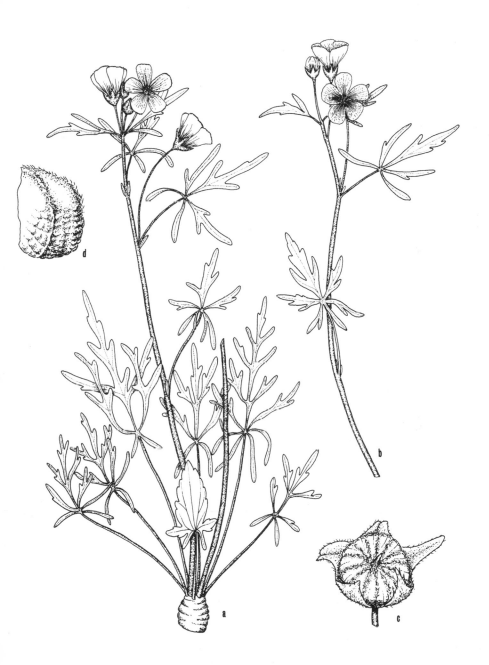

11. Callirhoë alcaeoides (Pale Poppy Mallow). *a.* Habit, × ½. *b.* Leafy branch, with flowers, × 1. *c.* Fruiting carpels, × 5. *d.* Segment of fruit, × 10.

cence is also unique among the Illinois species of *Callirhoë*. This is one of the rarer native plants in Illinois. It was first collected in Peoria County in July, 1897, by F. McDonald on dry, gravelly knolls at Peoria. Then, on May 30, 1950, Virginius H. Chase collected it from a gravel slope at the Horseshoe Bottom overlook. Both these collections undoubtedly represent specimens from native populations.

Specimens collected along railroads in Christian and Henry counties probably are of adventive plants.

This species flowers from late May to mid-August.

4. **Callirhoë digitata** Nutt. Journ. Acad. Phil. 2:181. 1821. *Fig. 12*.

Perennial herb from a thick, woody root; stems erect, branched from the base, to 1 m tall, glabrous or sparsely hirsute only at base, glaucous; basal leaves triangular to orbicular, cordate at the base, palmately lobed or divided, mostly glabrous, to 8 cm long, the petioles glabrous; upper leaves palmately divided, often nearly to the base, the segments linear, usually entire; flowers terminal, 3.0–4.5 cm broad, on slender, glabrous pedicels, without involucral bracts; sepals 5, united at the base, the lobes triangular to lanceolate, 6–10 mm long; petals 5, free, cuneate, shallowly several-notched at the apex, reddish-purple, 2–3 cm long; fruiting carpels numerous, scarcely beaked, minutely pubescent to glabrous, rugose-reticulate.

COMMON NAME: Fringed Poppy Mallow.

HABITAT: Dry fields.

RANGE: Native to the western United States; rarely adventive east of the Mississippi River.

ILLINOIS DISTRIBUTION: Known only from DuPage, Henderson, and Kane counties.

This very rare adventive species lacks an involucre of bracts below the calyx, just as in *C. alcaeoides*, but has larger, reddish-purple flowers.

The fringed poppy mallow, so-named because of the short fringe at the tip of each petal, flowers from late May to July.

3. Iliamna Greene

Herbs from rhizomes; leaves alternate, palmately lobed; flowers axillary or terminal, solitary or clustered, perfect, actinomorphic, with an involucre of bracts; sepals 5, united at the base; petals 5,

12. *Callirhoë digitata* (Fringed Poppy Mallow). *a*. Flowering branch, ×¾. *b*. Leaf, ×1. *c*. Flower, ×1½. *d*. Fruiting carpels, ×5. *e,f*. Segment of fruit, ×10.

free; stamen column anther-bearing only at the tip; pistil several-carpellate, each carpel 2–4 ovulate; fruiting carpels arranged in a circle, dehiscent, 2- to 4-seeded.

Iliamna is composed of eight species, all but two native to the western United States. Only our species and *I. corei* Sherff of Virginia are native east of the Mississippi River.

Only the following species occurs in Illinois.

1. **Iliamna remota** Greene, Leaflets 1:206. 1906. *Fig. 13.*
Sphaeralcea remota (Greene) Fern. Rhodora 10:52. 1908.
Phymosia remota (Greene) Britt. in Britt. & Brown, Ill. Fl. ed.
2, 2:522. 1913.
Iliamna remota Greene var. *typica* Sherff, Rhodora 48:93. 1946.

Herbaceous perennial; stems erect, much branched, to nearly 2 m tall, stellate-pubescent; leaves orbicular to reniform in outline, cordate, palmately 5- to 7-lobed, the lobes triangular, acute, dentate, stellate-pubescent on both surfaces, to 20 cm broad, the petioles stellate-pubescent, to 15 cm long; stipules subulate, to 10 mm long; flowers in small axillary clusters, 2.5–5.0 cm broad, the pedicels to 1 cm long, the involucral bracts linear, 5–10 mm long; sepals 5, united at the base, stellate-pubescent, 12–18 mm long, the lobes broadly lanceolate, 8–10 mm long, enlarging in fruit; petals obovate, emarginate, rose, 2.5–3.0 cm long; fruiting carpels numerous, arranged in a circle, oblongoid, stiffly pubescent on the back; seeds reniform, 3 mm high, brown, densely pubescent.

COMMON NAME: Kankakee Mallow.
HABITAT: Gravelly soil.
RANGE: Known only in the native condition from Kankakee County, Illinois.
ILLINOIS DISTRIBUTION: Altorf Island, Kankakee County. As interpreted in this work, the Kankakee mallow is known in a native condition only from Altorf Island in the middle of the Kankakee River, where it was first found by E. J. Hill on June 29, 1872. This species still occurs on the island, where it grows with *Carya cordiformis, Celtis occidentalis, Prunus serotina, Quercus macrocarpa, Campanula americana,* and *Sanicula canadensis.*

Adventive localities for this species have been found in Indiana and West Virginia.

There is a similar plant in the vicinity of The Narrows, Giles County, Virginia, which some botanists equate with our species, but since the geographic isolation is so great, and because there are some morphological differences, I prefer to consider the Virginia plant as a separate species, *I. corei* Sherff.

The Kankakee mallow flowers from late June to mid-August.

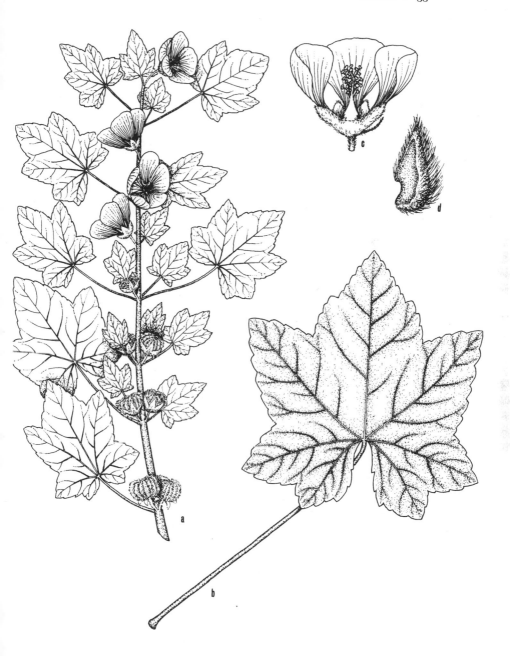

13. *Iliamna remota* (Kankakee Mallow). *a*. Upper part of plant, × ½. *b*. Lower leaf, × ½. *c*. Flower, × 1. *d*. Section of fruit, × 2½.

4. Gossypium L.–Cotton

Annual or perennial herbs or becoming treelike; leaves alternate, palmately veined or lobed, often glandular-punctate; flowers axillary, perfect, actinomorphic, with an involucre of 3–7 free or united bracts; sepals 5, united; petals 5, free; stamen column anther-bearing only at the tip; ovary 3- to 5-locular, with 2–7 ovules per locule; fruit a dehiscent capsule with white-woolly seeds.

Gossypium is a genus of about two dozen species native to most tropical regions of the World. Only a few of the species are cotton-producing.

In addition to the following escaped species, Gossypium barbadense L. is another important cotton producer.

1. Gossypium hirsutum L. Sp. Pl. ed. 2, 975. 1763. Fig. 14.

Stout annual herb; stems erect, much branched, to 1.3 m tall, more or less hirsute; leaves 3-lobed, acuminate at the apex, cordate to truncate at the base, more or less hirsute, to 15 cm long, nearly as broad; flowers axillary, to 6.5 cm broad, pedicellate, the involucral bracts deeply lobed, more than half as long as the corolla; sepals 5, united below, green; petals 5, free, white or pale yellow, turning pink or purple, to 3.5 cm long; fruits to 5 cm long, containing 8–10 seeds, each seed covered with white fibers.

COMMON NAME: Cotton.

HABITAT: Cultivated fields.

RANGE: Native of tropical America; uncommonly adventive in the United States.

ILLINOIS DISTRIBUTION: Adventive in Alexander, Champaign, and Cook counties.

Cotton was grown as a commercial crop, particularly in Alexander County, for several years. A cotton gin operated near Cairo until about 1970.

Cotton is the most important plant fiber used by man to make clothing. The boll, which contains the cotton fibers, forms as the flower withers. Within each boll, at maturity, are usually five groups, called locks, each composed of about 10 seeds to which the fibers are attached.

Cotton is also valuable in the production of cottonseed oil. The oil, which is derived from the seeds, is used in the making of salad oils, shortening, margarine, and soap.

Cotton flowers during the summer.

14. Gossypium hirsutum (Cotton). *a.* Upper part of plant, with flower and fruit, × ½. *b.* Leaf, × 1. *c.* Flower, × ½.

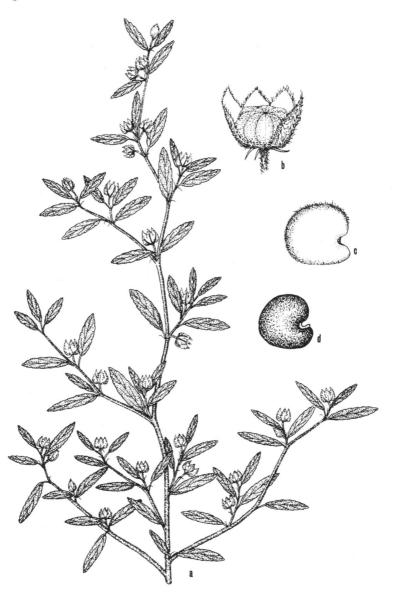

15. *Sphaeralcea angusta* (Globe Mallow). *a.* Upper part of plant, × ½. *b.* Fruit, with one sepal removed, × 2½. *c.* Fruiting carpel, × 5. *d.* Seed, × 5.

5. *Sphaeralcea* St. Hil.–Globe Mallow

Annual or perennial herbs; leaves alternate, entire to lobed; flowers axillary or terminal, solitary or in racemes, perfect, actinomorphic, with an involucel of 1–3 bracts, or involucel absent; sepals 5, united; petals 5, free; stamen column anther-bearing only at the tip; ovary with 5 or more locules, with 1 ovule per locule; fruit indehiscent or 2-valved, sometimes beaked.

Sphaeralcea is a genus of about seventy-five species native to North and South America and South Africa.

Only the following species occurs in Illinois.

1. **Sphaeralcea angusta** (Gray) Fern. Rhodora 41:435. 1939. *Fig. 15.*

Malvastrum angustum Gray, Mem. Am. Acad. II. 4:21. 1849.

Annual herb from slender roots; stems erect, branched, appressed-pubescent, to 50 cm tall; leaves linear-oblong to oblong-lanceolate, acute at the apex, cuneate at the base, dentate, to 2 cm long, to 9 mm broad, appressed-pubescent; flower solitary from the upper axils, to 1.2 cm broad, on pedicels as long as or slightly longer than the flowers, the bracts linear, shorter than the calyx; sepals 5, united at the base, ovate-deltoid, pubescent, 5–8 mm long, inflated; petals 5, free, rarely absent, yellow, about the same length as the sepals; fruiting carpels 5, reniform, glabrous or puberulent, 2-valved at maturity.

COMMON NAME: Globe Mallow.

HABITAT: Dry soil.

RANGE: Illinois to Nebraska, south to Kansas, Missouri, and Alabama.

ILLINOIS DISTRIBUTION: Known only from Grundy, LaSalle, Rock Island, and Will counties.

There is a superficial resemblance between this species and *Sida spinosa*. However, in *Sida spinosa*, there are no involucral bracts subtending the calyx.

Several authors consider this species to belong to the genus *Malvastrum*, where Asa Gray originally described it.

The Grundy County station is at Goose Lake Prairie where the globe mallow occurs in degraded prairie.

The flowers bloom in July and August.

6. *Hibiscus* L.–Rose Mallow

Herbs or shrubs or, in the tropics, small trees; leaves alternate, toothed or lobed; flowers usually large, showy, axillary or in panicles, perfect, actinomorphic, with an involucel of several bracts; sepals 5, united; petals 5, free; stamen column anther-bearing below along much of its length; ovary with 5 locules, with 3 or more ovules per locule, the style branches 5, the stigmas capitate; capsule dehiscent, 5-valved, with reniform seeds.

Hibiscus is a genus of nearly two hundred species native to warm temperature regions of the World. Most species have large, showy flowers. Many are prized as ornamentals, including *H. rosasinensis* L. (Chinese hibiscus), *H. mutabilis* L. (cotton rose), *H. palustris* L. (marsh mallow), and *H. syriacus* (rose-of-Sharon). *Hibiscus esculentus* L., okra or gumbo, is grown for its edible fruits.

KEY TO THE SPECIES OF *Hibiscus* IN ILLINOIS

1. Annual herbs; flowers yellow or whitish with a black or purple center _____ 2
1. Perennial herbs, shrubs, or small trees; flowers not yellow _____ 3
    2. Calyx inflated, 5-winged; seeds verrucose _____ 1. *H. trionum*
    2. Calyx like a spathe, splitting down one side; seeds mucilaginous _____ 2. *H. esculentus*
3. Stems and leaves glabrous _____ 4
3. Stems and at least the lower surface of the leaves pubescent _____ 5
    4. Herb 1.0–2.5 m tall; calyx glabrous or with simple hairs _____
    _____ 3. *H. militaris*
    4. Shrub or small tree 3 m tall or taller; calyx stellate-pubescent ____
    _____ 4. *H. syriacus*
5. Leaves glabrous or nearly so above; some of the leaves often 3-lobed; capsules glabrous _____ 5. *H. palustris*
5. Leaves soft-hairy on both surfaces; none of the leaves lobed; capsules densely hairy _____ 6. *H. lasiocarpos*

1. Hibiscus trionum L. Sp. Pl. 697. 1753. *Fig. 16*.

Annual from fibrous roots; stems spreading to ascending, branched from the base, with spreading pubescence; leaves palmately 3- to 7-lobed, the middle lobe much the longest, all lobes obtuse, dentate, pubescent, the petioles with spreading pubescence; flowers from the uppermost leaf axils, to 5.5 cm across, yellow with a purple center, on spreading-pubescent pedicels, the bracts linear, ciliate, much shorter than the calyx; calyx inflated, 5-angled, membranous,

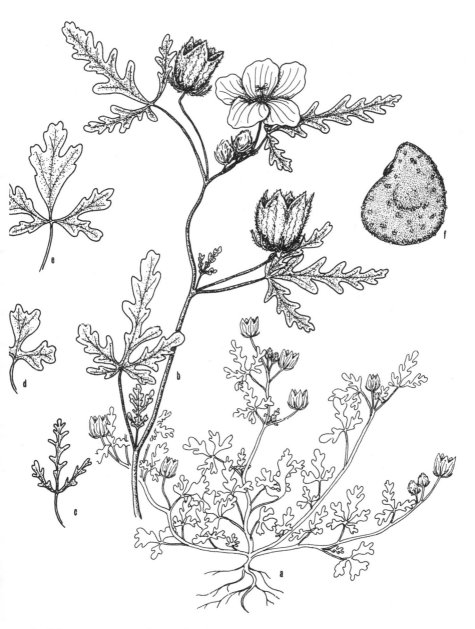

16. *Hibiscus trionum* (Flower-of-an-hour). *a.* Habit, × ⅙. *b.* Leafy branch, with flower and fruits, × ½. *c,d,e.* Leaf variations, × ½. *f.* Seed, × 10.

the dark nerves hispid; petals 5, orbicular to obovate, to 2.8 cm long, many-nerved; capsule globose to ovoid, pubescent, with many rough, verrucose seeds.

COMMON NAME: Flower-of-an-hour.
HABITAT: Fields, pastures, roadsides, waste ground.
RANGE: Native of southern Europe; adventive in most of North America.
ILLINOIS DISTRIBUTION: Occasional to common in the northern three-fourths of the state, rare in the southern one-fourth.
Flower-of-an-hour receives its common name from the fact that each flower is open for only a very short time. It is totally different from the genus *Talinum* of the Portulacaceae, which also is called flower-of-an-hour.

This species differs from all others in the genus except *H. esculentus* by its yellow flowers and annual habit. From *H. esculentus* it differs by its inflated calyx and its globose or ovoid capsules with rough, verrucose seeds.

June to October is the time for flowering for this species.

2. Hibiscus esculentus L. Sp. Pl. 693. 1753. *Fig. 17.*

Annual from a tuft of roots; stems ascending, to 2 m tall, branched, glabrous; leaves ovate in outline, to 30 cm across, 3-to 9-lobed, cordate at the base, the lobes coarsely toothed; flowers from the uppermost axils, yellow with a reddish center, to 6 cm across, with linear bracts to 2 cm long; calyx spathaceous, split down one side, falling away early; petals 5, orbicular to obovate, to 3 cm long, many-nerved; capsule cylindrical, beaked, strongly ribbed, to 15 cm long, becoming woody at maturity, with numerous mucilaginous seeds.

COMMON NAMES: Okra, Gumbo.
HABITAT: Along a railroad.
RANGE: Native of Africa; occasionally cultivated in North America but rarely escaped.
ILLINOIS DISTRIBUTION: Known only from Champaign County.
Okra is sometimes cultivated in vegetable gardens for its edible fruits. The combination of yellow flowers and glabrous stems and leaves distinguishes it from all other species of *Hibiscus* in Illinois.
The large flowers bloom from July to September.

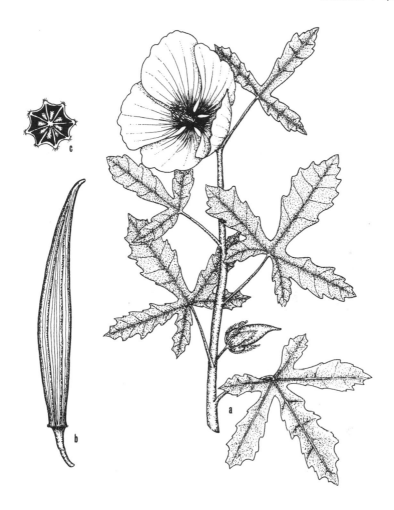

17. Hibiscus esculentus (Okra). *a.* Upper part of plant, × ½. *b.* Fruit, × ½. *c.* Cross-section of fruit, × 1.

3. Hibiscus militaris Cav. Diss. 3:352, pl. 198, f. 2. 1787. *Fig. 18.*

Perennial from stout rootstocks; stems erect, to 2.5 m tall, glabrous; leaves ovate, acute to acuminate at the apex, cordate or truncate at the base, to 10 cm long, serrate, some of the leaves hastately 3-lobed, glabrous, on glabrous petioles; flowers from the uppermost axils, pink with a purple center, rarely white, to 7.5 cm long, on glabrous peduncles, to 10 cm long, with linear bracts about as long

18. Hibiscus militaris (Halberd-leaved Rose Mallow). *a*. Leafy branch, with flower and fruits, × ½. *b*. Fruit, × 1. *c*. Seed, × 5.

as the calyx; calyx 5-lobed at the tip, 2–3 cm long, glabrous or nearly so, striate-nerved, becoming inflated in fruit; petals obovate, striate-nerved, to 7 cm long; capsule ovoid, glabrous, with silky-pubescent seeds.

COMMON NAME: Halberd-leaved Rose Mallow.

HABITAT: Wet areas.

RANGE: Pennsylvania to Minnesota and Nebraska, south to Texas and Florida.

ILLINOIS DISTRIBUTION: Occasional throughout the state.

This handsome species of wet ground usually has pink flowers with a dark purple center. Occasional white-flowered specimens occur. One colony of such plants grows along Hay-glade Ditch one-half mile west of Ware, in Union County.

The glabrous leaves distinguish it from *H. palustris* and *H. lasiocarpos*, two other tall perennial herbs in Illinois.

Some or all the leaves on the plants are hastately 3-lobed.

The silky-hairy seeds are eaten by various kinds of wild birds.

The flowers occur from July to October.

4. Hibiscus syriacus L. Sp. Pl. 695. 1753. *Fig. 19.*

Shrub to 6 m tall; stems erect, branched, glabrous or nearly so; leaves ovate, to 8 cm long, acute to subacute at the apex, rounded to cuneate at the base, unlobed or 3- to 5-lobed, dentate, glabrous or stellate-pubescent above, on usually glabrous petioles; flowers from the uppermost axils, pink or white, often with a crimson center, to 8 cm across, subtended by stellate-pubescent, linear bracts about as long as the calyx; calyx 5-lobed at the tip, 2–3 cm long, stellate-pubescent; petals 5, orbicular to broadly obovate, to 7.5 cm long; capsule ovoid, to 2.5 cm long.

COMMON NAMES: Rose-of-Sharon; Shrubby Althaea.

HABITAT: Along an abandoned rural road.

RANGE: Native of eastern Asia; rarely escaped from cultivation in North America.

ILLINOIS DISTRIBUTION: Jackson Co.: Little Grassy camp, Southern Illinois University, August 14, 1969, *R. H. Mohlenbrock s.n.*

Rose-of-Sharon is frequently found growing around farm homes. Rarely does it escape from cultivation.

19. Hibiscus syriacus (Rose-of-Sharon). *a.* Leafy branch, with flower, × ¾. *b.* Outside of flower, × ¾.

Numerous color forms of the flowers are known, and some "double-flowered" varieties are known.
The flowers boom from July to September.

5. **Hibiscus palustris** L. Sp. Pl. 693. 1753. *Fig. 20.*
Perennial herb from stout rootstocks; stems erect, to 2.5 m tall, branched, canescent; leaves ovate to ovate-lanceolate, acute to acuminate at the apex, rounded to cordate at the base, to 15 cm long, dentate, occasionally 3-lobed, dark green and glabrous or nearly so above, canescent below, the petioles to 10 cm long; flowers clustered near the tips of the stems, to 15 cm across, pink, rose, or purplish, usually crimson in the center, the peduncles often adnate to the petioles, the bracts linear, glabrous, shorter than the calyx; calyx 5-lobed at the tip, to 3 cm long, the lobes ovate; petals 5, obovate, several-nerved, to 10 cm long; capsule globose, up to 2.5 cm across, glabrous, with numerous glabrous seeds.

COMMON NAME: Swamp Rose Mallow.
HABITAT: Marshes and other wet soil.
RANGE: New York and southern Ontario to Michigan, northern Indiana, and northern Illinois; Massachusetts to North Carolina.
ILLINOIS DISTRIBUTION: Scattered throughout the state, but nowhere common.
This species has often been called *H. moscheutos* L., but this latter binomial refers to a different species of the eastern United States.
The flowers bloom from July to September.

6. **Hibiscus lasiocarpos** Cav. Diss. 3:159, pl. 70, f. 1. 1787. *Fig. 21.*
Perennial herb from stout rootstocks; stems erect, to 2 m tall, branched, pubescent; leaves ovate, acute to acuminate at the apex, cordate to rounded at the base, to 15 cm long, dentate, unlobed or occasionally 3- to 7-lobed, softly stellate-pubescent on both surfaces, the petioles to 10 cm long; flowers from the uppermost axils, to 10 cm across, white or rose, with a crimson center, the bracts linear, ciliate, usually as long as the calyx; calyx 5-lobed, pubescent, to 3 cm long; petals 5, obovate, several-nerved, to 10 cm long; capsule ovoid, densely villous, to 2.5 cm long, with numerous glabrous or finely pubescent seeds.

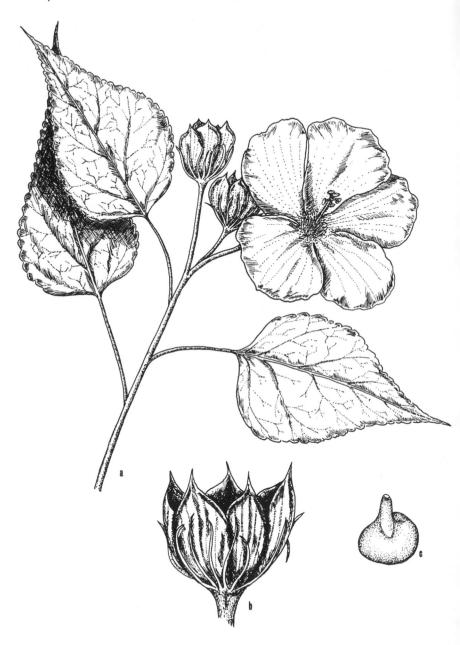

20. *Hibiscus palustris* (Swamp Rose Mallow). *a*. Leafy branch, with flower and fruits, × ⅓. *b*. Fruit, × 1½. *c*. Seed, × 5.

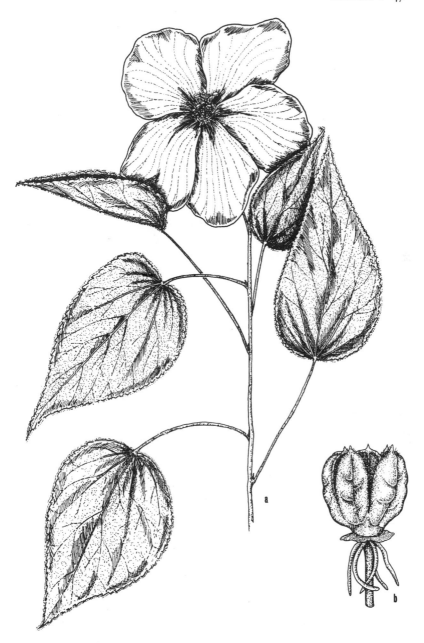

21. *Hibiscus lasiocarpos* (Hairy Rose Mallow). *a*. Leafy branch, with flower, × ⅓. *b*. Fruit, × 1½.

COMMON NAME: Hairy Rose Mallow.

HABITAT: Wet ground around ponds, in swamps, and in woods.

RANGE: Indiana to Missouri, south to Texas and Georgia.

ILLINOIS DISTRIBUTION: Occasional to common in the southern half of the state; also in McLean County.

The hairy rose mallow is the most common species of *Hibiscus* in the southern half of the state where it grows in wet woods, swamps, and around ponds and lakes.

This is the only species of *Hibiscus* in Illinois with the leaves softly hairy on both surfaces.

The flowers range in color from white to rose. They appear from July to October.

7. Althaea L.

Annual, biennal, or perennial herbs; leaves alternate, usually lobed; flowers usually large, showy, axillary, solitary or in racemes, perfect, actinomorphic, with an involucel of 6–9 bracts connate at the base; sepals 5, united; petals 5, free; stamen column anther-bearing only at the tip; pistil several-carpellate, each carpel 1-ovulate; fruiting carpels arranged in a circle, indehiscent, 1-seeded, beakless.

Althaea is a genus of about fifteen showy species native to the Old World. In technical characters, it is closely related to *Malva*, differing by having 6–9 involucel bracts per flower, rather than 0–3.

Only the following species is known from Illinois.

1. Althaea rosea (L.) Cav. Diss. 2:91, t. 29, f. 3. 1790. *Fig. 22*.

Alcea rosea L. Sp. Pl. 687. 1753.

Biennial herb; stems erect, more or less branched, to 3 m tall, rough-pubescent; leaves ovate to nearly orbicular in outline, 5- to 7-lobed or -angled, cordate at the base, crenate, rough-pubescent, rugose; flowers nearly sessile in an elongated raceme or spike, variously colored, to 8 cm across, each flower subtended by 6–9 narrow, basally connate bracts; sepals green, pubescent, to 2 cm long, united at the base; petals 5, free, oblong-spatulate, to 7 cm long; carpels several, arranged in a circle, more or less pubescent.

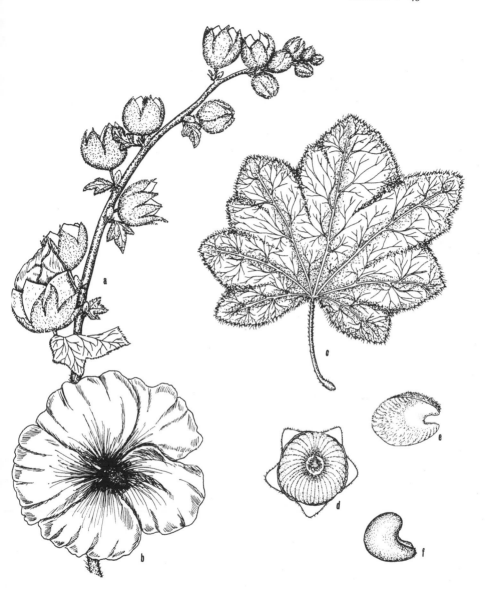

22. *Althaea rosea* (Hollyhock). *a*. Leafy branch, with buds, ×½. *b*. Flower, ×⅓. *c*. Leaf, ×½. *d*. Fruit, ×5. *e,f*. Seed, ×10.

COMMON NAME: Hollyhock.

HABITAT: Waste ground, particularly along roads and railroads.

RANGE: Native of Europe and Asia; occasionally escaped from cultivation throughout much of the United States.

ILLINOIS DISTRIBUTION: Probably in almost every Illinois county, but collected only in a few.

Hollyhock is an old-fashioned garden favorite which frequently is found growing along roads and railroads.

The flowers, which bloom from July to October, occur in a variety of colors.

8. Napaea L.–Glade Mallow

Perennial, dioecious herbs; leaves alternate, palmately lobed; flowers unisexual, in terminal corymbose panicles, without involucels; calyx 5-lobed; petals 5, free; stamen column anther-bearing only at the tip; pistillate flowers with 8–10 styles from the tip of a sterile stamen column, the pistils 8- to 10-carpellate, each carpel 1-ovulate; fruiting carpels imperfectly 2-valved, beakless, 1-seeded.

Only the following species comprises the genus.

1. Napaea dioica L. Sp. Pl. 686. 1753. *Fig. 23.*

Sida dioica (L.) Cav. Diss. 5:278, t. 132. 1788.

Perennial herb; stems erect, simple or branched, to 3 m tall, strigose to nearly glabrous; basal and lower leaves deeply 7- to 11-parted, cordate at the base, the lobes acute, dentate or incised, pubescent at least when young, on long petioles; upper leaves less deeply 5- to 9-lobed, the lobes acuminate, sharply toothed, on short petioles; staminate flowers white, to 2 cm broad; pistillate flowers white, to 1.7 cm broad; sepals 5, lanceolate, united at the base, to 1 cm long; petals 5, free, 2.0–2.5 cm long, obovate; fruiting carpels reniform, 1-seeded, 1-nerved, rugose.

COMMON NAME: Glade Mallow.

HABITAT: Alluvial soil along streams and rivers.

RANGE: Pennsylvania to Minnesota, south to Iowa, central Illinois, and Virginia.

ILLINOIS DISTRIBUTION: Occasional in the northern half of the state, apparently absent in the southern half, except for Clark County.

23. *Napaea dioica* (Glade Mallow). *a.* Flowering branch, ×¼. *b.* Leaf, ×½. *c.* Staminate flower, ×2½. *d.* Pistillate flower, ×2½. *e.* Fruiting carpels, ×2½. *f.* Fruiting carpel, ×5. *g.* Seed, ×5.

This species, which has a restricted range, is of local occurrence in the northern half of the state where it grows in alluvial soils along streams and rivers. Swink (1974) lists some of its associates as *Amorpha fruticosa, Campanula americana, Eupatorium rugosum, Rudbeckia laciniata*, and *Silphium perfoliatum*. The flowers appear in June and July.

9. Anoda Cav.

Annual herbs; leaves alternate, usually lobed; flowers solitary from the uppermost axils, without a subtending involucel; calyx 5-lobed; petals 5, free; staminal column anther-bearing at the apex; pistils several-carpellate, each carpel 1-ovulate; fruiting carpels several in a depressed capsule, the lateral walls becoming absorbed with age.

This genus of about ten annual herbs is native from the southern United States into South America.

Only the following species occurs in Illinois.

1. Anoda cristata (L.) Schlecht. Linnaea 11:210. 1837. *Fig. 24.*

Sida cristata L. Sp. Pl. 685. 1753.

Annual herb from tufted roots; stems erect, branched, to 75 cm tall, villous-hirsute; leaves ovate to deltoid, unlobed to hastate to several-lobed, coarsely crenate, pubescent, to 5 cm long, the petioles to 3 cm long; flowers solitary from the upper axils, pale blue to purple, to 4 cm across, the pedicels filiform, to 7 cm long; calyx 5-parted, the lobes lanceolate-ovate, acuminate, green, becoming spread out flat beneath the fruiting carpels; petals 5, free, to 2.5 cm long; fruiting carpels 9–20, awned, hispid, the seeds minutely pubescent.

COMMON NAME: Crested Anoda.

HABITAT: Waste ground.

RANGE: Native to the southwestern United States and Mexico; occasionally adventive in the eastern United States.

ILLINOIS DISTRIBUTION: Known only from Hancock and Massac counties.

S. B. Mead collected this species in the mid-nineteenth century several times near Augusta in the vicinity of gardens from which it undoubtedly escaped. John Schwegman discovered this species in Massac County in 1974.

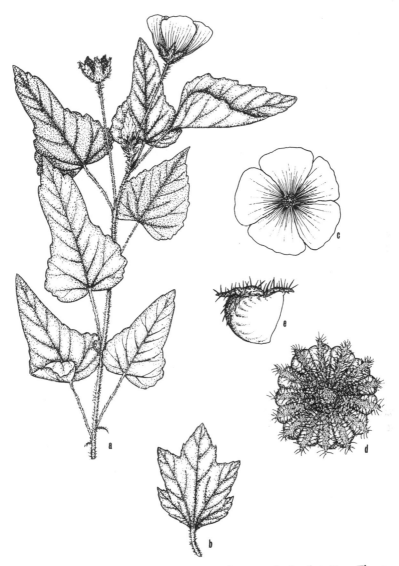

24. *Anoda cristata* (Crested Anoda). *a*. Habit, ×1. *b*. Leaf, ×½. *c* Flower, ×1. *d*. Fruiting carpels, ×2.½ *e*. Fruiting carpel, ×5.

Anoda cristata may be distinguished from all other members of the Malvaceae by its bisexual flowers up to 4 cm across and the absence of an involucel of bracts.

The flowers are borne from August to October.

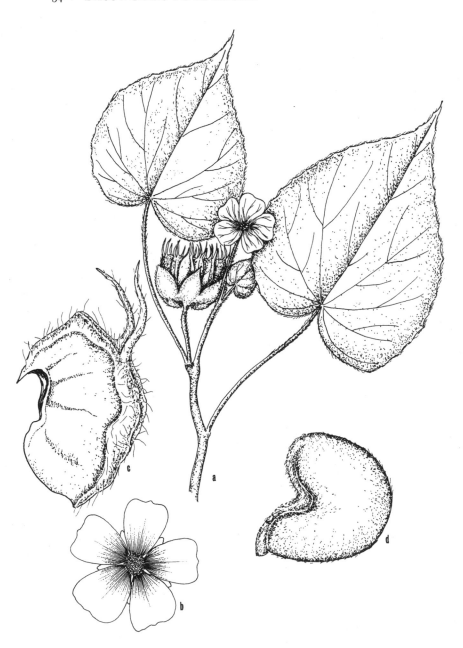

25. *Abutilon theophrastii* (Velvet Leaf). *a*. Leafy branch, with flower and fruit, × ¼. *b*. Flower, × 1½. *c*. Fruiting carpel, × 2. *d*. Seed, × 2.

10. Abutilon Mill.–Indian Mallow

Herbs, shrubs, or trees (in the tropics); leaves alternate, often lobed or angled; flowers from the uppermost axils, without a subtending involucel; calyx 5-lobed; petals 5, free; staminal column anther-bearing at the apex; pistils with 5 or more carpels, each carpel 2- to 9-ovulate; fruiting carpels 2-valved, beaked, falling away at maturity.

Abutilon is a genus of about one hundred species native to the warm temperate and tropical regions of the World. Abutilon technically is most closely related to the genus Sida, from which it differs by having two or more seeds per carpel. Several species are grown as garden ornamentals.

Only the following species occurs in Illinois.

1. Abutilon theophrastii Medic. Malv. 28. 1787. Fig. 25.

Sida abutilon L. Sp. Pl. 685. 1753.
Abutilon avicennae Gaertn. Fruct. & Sem. 2:251, pl. 135. 1791.
Abutilon abutilon (L.) Rusby, Mem. Torrey Club 5:222. 1894.

Stout annual herb from tufted roots; stems erect, branched, to 1.5 m tall, velvety-pubescent; leaves ovate to ovate-orbicular, acuminate at the apex but with a blunt tip, cordate at the base, entire or denticulate, velvety-pubescent, to 30 cm long, nearly as broad, the petioles to 20 cm long, velvety-pubescent; flowers solitary, axillary, to 2 cm across, on stout peduncles shorter than the petioles; calyx 5-parted more than half way to the base, the lobes ovate or lance-ovate, green, pubescent, to 1 cm long; petals 5, free, yellow, obspatulate, nerved, to 1.5 cm long, fruiting head 2–4 cm broad, the carpels 12–15, pubescent, slender-beaked, dehiscing from the apex.

COMMON NAMES: Velvet Leaf; Indian Mallow; Butterprint.

HABITAT: Waste places, often in fields.

RANGE: Native of India; widely adventive in North America.

ILLINOIS DISTRIBUTION: Occasional to common throughout the state.

Abutilon theophrastii is a stout, fast-growing annual of fields and barnyards. The way in which the 12–15 carpels are arranged in the fruiting head is reminiscent of the print blocks used by farmers to stamp butter rolls.

In northern Illinois, this species is commonly associated with *Hibiscus trionum*. Another common associate throughout the state is *Datura stramonium*.

Until 1912, this species was known by Illinois botanists as *A. avicennae*.

The yellow flowers bloom from June to October.

11. Sida L.

Annual or perennial herbs; leaves alternate, toothed or lobed; flowers solitary or clustered from the upper axils or terminal, without a subtending involucel; calyx 5-lobed; petals 5, free; staminal column anther-bearing at the apex; pistils with 5 to many carpels, each carpel 1-ovulate; fruiting carpels indehiscent or tardily dehiscent along 2 valves, each with a pendulous seed.

Sida is a genus of about one hundred species native to warm temperate and tropical regions of the World.

KEY TO THE SPECIES OF Sida IN ILLINOIS

1. Leaves oblong to lance-ovate, all or most of them over 8 mm broad; carpels 5 per flower _____ 1. *S. spinosa*
1. Leaves linear, rarely over 7 mm broad; carpels 8 or more per flower _____ 2. *S. elliottii*

1. Sida spinosa L. Sp. Pl. 683. 1753. *Fig. 26*.

Annual from tufted roots; stems erect, branched, to 65 cm tall, softly pubescent; leaves oblong to lance-ovate, obtuse to acute at the apex, rounded to cordate at the base, serrate, to 4 cm long, usually at least 8 mm broad, finely pubescent on petiole to 2 cm long, often with a short, spinose tubercle at the base; flowers axillary, solitary, to 1 cm across, on peduncles shorter than the petioles; calyx 5-parted to about the middle, the lobes deltoid, acute, green; petals 5, free, yellow, to 7 mm long; fruiting carpels 5, loosely united into an ovoid fruit, each carpel dehiscing at the apex into two beaks.

COMMON NAME: Prickly Sida.
HABITAT: Fields, roadsides, along railroads.
RANGE: Native of tropical America; adventive in much of the eastern half of the United States.
ILLINOIS DISTRIBUTION: Common in the southern four-fifths of the state, becoming uncommon in the northern one-fifth and apparently absent in the northwestern counties.

26. *Sida spinosa* (Prickly Sida). *a*. Habit, ×½. *b*. Fruiting cluster, ×5. *c*. Single segment of fruiting cluster, ×10.

Prickly sida derives its common name from the sometimes sharp-pointed tubercle at the base of the petioles of some of the upper leaves.

This species is common throughout much of Illinois, but is often overlooked because of its inconspicuous nature. Its broader leaves and five carpels distinguish it from the much taller *S. elliottii*.

Mead reported this species from Illinois as early as 1846.
Prickly sida blooms from June to October.

2. Sida elliottii Torr. & Gray, Fl. N. Am. 1:231. 1838. *Fig. 27*.

Perennial herb; stems erect, branched, to about 1 m tall, glabrous
or nearly so; leaves linear, acute at the apex, cuneate at the base,
serrulate, glabrous or nearly so, to 4.5 cm long, rarely over 7 mm
broad, the petioles up to 8 mm long, without spinose tubercles at
the base; flowers solitary and axillary, on peduncles longer than the
petioles; calyx 5-parted to about the middle, the lobes deltoid,
acute, green, strigose or nearly glabrous; petals 5, free, yellow, to
1.5 cm long; fruiting carpels 8 or more, loosely united into an ovoid
fruit, each carpel dehiscing at the apex into two beaks.

COMMON NAME: Elliott's Sida.

HABITAT: Farm lot.

RANGE: South Carolina to southeastern Missouri, south
to Alabama and Florida.

ILLINOIS DISTRIBUTION: Known only from Alexander
County.

This rare species, first discovered in Illinois in September, 1971, was found growing by the author in a sandy
farm lot.

The plants are much taller than those of *S. spinosa* and
lack the spinose tubercles at the base of the uppermost
petioles. The leaves are also narrower than those of *S. spinosa*, and
there are eight or more carpels per flower.

The yellow flowers bloom from August to October.

27. *Sida elliottii* (Elliott's Sida). *a*. Leafy branch, with flowers, ×1. *b*. Fruiting carpels, ×5. *c*. Fruiting carpel, ×10.

Order Urticales

In the Thorne system of classification (1968) followed here, the Urticales is the second of six orders comprising the superorder Malviiflorae. There is considerable departure among botanists from this scheme. For example, Cronquist (1968) places the Urticales in his subclass Hamamelidae, along with eight other orders. It is interesting to note that the eight orders Cronquist groups with the Urticales in his Hamamelidae are completely different from the five orders Thorne groups with the Urticales in his Malviiflorae. Engler generally believed that the Urticales were closest to the Fagales, while Bessey incorporated the families of the Urticales in the Malvales.

In any event, almost all botanists have attributed the same families to the Urticales. In Illinois this includes the Ulmaceae, the Moraceae (including the Cannabinaceae), and the Urticaceae.

The unifying characters of the Urticales are the bicarpellate unilocular superior ovary, each with one ovule.

I am following Thorne in relegating the sometimes recognized family Cannabinaceae as a subfamily of the Moraceae.

As treated here, the Urticales in Illinois is composed of three families, thirteen genera, and twenty-five species.

ULMACEAE–ELM FAMILY

Monoecious trees or shrubs; leaves simple, alternate, with caducous stipules; flowers perfect or unisexual, zygomorphic, solitary or variously arranged; sepals 4–8, free or united at the base; petals absent; stamens 4–8, free, arising from the hypanthium; pistil one, the ovary superior, bicarpellate, unilocular, with one pendulous ovule per locule, the styles 2; fruit a samara, nut, or drupe.

The Ulmaceae are composed of fifteen genera and about 175 species. Most of these are native to the Northern Hemisphere. There are three genera and ten species in Illinois. Species of *Ulmus* and *Celtis* are grown as ornamentals.

The family is divided into two tribes. Tribe Ulmeae, which includes *Ulmus* and *Planera*, has united sepals, solid pith, and a sa-

mara or nutlike fruit. Tribe Celtideae, including *Celtis*, has free
sepals, chambered pith, and a drupe.

KEY TO THE GENERA OF Ulmaceae IN ILLINOIS

1. Leaves with strong lateral veins arising from the main vein at a distance
   above the very base of the blade; pith of branches solid; sepals united;
   fruit a samara or nut _____ 2
1. Leaves with a pair of strong lateral veins arising from the main vein at
   the very base of the blade; pith of branches chambered; sepals free, or
   nearly so; fruit a drupe _____ 3. *Celtis*
   2. Fruit a samara; all flowers perfect, appearing before the leaves;
      leaves mostly doubly toothed _____ 1. *Ulmus*
   2. Fruit a wingless nut; at least some of the flowers unisexual, appear-
      ing with the leaves; leaves mostly singly toothed ____ 2. *Planera*

1. *Ulmus* L.–Elm

Trees with furrowed bark; twigs often flexuous, sometimes winged,
with small, solid pith; bud scales 2-ranked; leaves alternate, simple,
serrate, the stipules linear to lanceolate, caducous; flowers mostly
perfect, clustered in fascicles, cymes, or racemes, borne on twigs of
the preceding year; calyx campanulate, 4- to 9-lobed; petals absent;
stamens 3–9, free, exserted beyond the calyx; pistil one, the ovary
superior, sessile or stipitate, 1- to 2-locular, with 1 ovule per locule,
the styles deeply 2-lobed; fruit a 1-seeded samara.

There are approximately eighteen species of *Ulmus* native to the
Northern Hemisphere. Four native and two introduced species are
known from Illinois.

In addition to the species enumerated below, *Ulmus parviflora*
Jacq. (Chinese elm) and *U. glabra* Huds. (Scotch elm) are sometimes grown as ornamentals.

KEY TO THE SPECIES OF Ulmus IN ILLINOIS

1. Upper surface of leaves harshly pubescent; winter buds rusty-pubescent; samaras eciliate on the margins but pubescent at the center of both
   sides _____ 1. *U. rubra*
1. Upper surface of leaves glabrous or pubescent, but never harshly so;
   winter buds glabrous or pubescent, but not with rusty pubescence; samaras ciliate or, if eciliate, the center of each side glabrous _____ 2
   2. None of the branches corky-winged _____ 3
   2. Some of the branches corky-winged _____ 4
3. Leaves doubly serrate, usually strongly asymmetrical at the base; samaras ciliate; flowers pendulous from long pedicels   2. *U. americana*

3. Leaves mostly singly serrate, usually nearly symmetrical at the base; samaras eciliate; flowers sessile or nearly so _____ 3. *U. pumila*
  4. Buds glabrous or nearly so; leaves sessile or on petioles up to 3 mm long; samaras (excluding cilia) up to 5 mm broad ____ 4. *U. alata*
  4. Buds downy-pubescent; leaves on petioles 3 mm long or longer; samaras (excluding cilia) 1.0–1.5 cm broad _____ 5
5. Most of the mature leaves over 8 cm long, glabrous above; flowers pendulous from long pedicels; samaras ciliate _____ 5. *U. thomasii*
5. Most of the mature leaves less than 8 cm long, scabrous above; flowers sessile or nearly so; samaras eciliate _____ 6. *U. procera*

1. **Ulmus rubra** Muhl. Trans. Am. Phil. Soc. 3:165. 1793. *Fig. 28.*

Ulmus fulva Michx. Fl. Bor. Am. 1:172. 1803.

Tree sometimes to 20 m tall, the crown broadly rounded, the trunk diameter up to 1.3 m; bark reddish-brown to gray, divided into flat ridges by shallow fissures; twigs moderately stout, reddish-brown, grayish-pubescent, the buds to 6 mm long, ovoid, obtuse, dark reddish-brown, rusty-pubescent; leaf scars crescent-shaped, slightly elevated, with 3 bundle scars; leaves ovate to ovate-oblong, acuminate at the apex, rounded at the very asymmetrical base, doubly serrate, dark green, rugose, and harshly scabrous on the upper surface, softly pubescent on the lower surface, sweetly fragrant when dry, to 15 cm long, up to half as broad, becoming dull yellow in the autumn, the petioles to 8 mm long, pubescent, the caducous stipules pubescent; flowers crowded in nearly sessile clusters, the pedicels softly pubescent, to 6 mm long, jointed near the base; calyx campanulate, to 4 mm long, 5- to 9-lobed, the lobes lanceolate, acute, green, pubescent; stamens 5–9, long-exserted, the filaments yellow, the anthers dark red; samaras orbicular to elliptic, to 2 cm across, subtended by the persistent calyx, terminated by the persistent, withered styles, the wings nerved, glabrous, except over the seed.

COMMON NAMES: Slippery Elm; Red Elm.
HABITAT: Woodlands, often in disturbed areas.
RANGE: Quebec to North Dakota, south to Texas and Florida.
ILLINOIS DISTRIBUTION: Common throughout the state; in every county.
Ulmus rubra is distinguished from all other elms in Illinois by its harshly pubescent leaves and its large, nonciliate samaras.

28. *Ulmus rubra* (Slippery Elm). *a.* Leafy branch, ×⅓. *b.* Staminate flowering cluster, ×1. *c.* Staminate flower, ×5. *d.* Fruiting clusters, ×½. *e.* Fruit, ×2.

Slippery elm is common as a small tree in many Illinois woods. It usually attains its greatest size in rich bottomland woods and along streams.

A piece of bark broken to show a cross-section shows a constant color, without alternating layers of color, as in the American elm. Slippery elm receives its name from the inner bark which becomes mucilaginous when chewed for a period of time.

The wood is heavy, hard, and tough. It is used for boats, boxes, baskets, barrel hoops, and woodenware, among other things.

For many years, this species was known as *Ulmus fulva*, but *U. rubra* clearly has priority as the correct binomial.

The flowers bloom from February to April, and the fruits develop when the leaves are about half grown.

2. Ulmus americana L. Sp. Pl. 226. 1753. *Fig. 29.*

Tree to 30 m tall, the crown broadly rounded, the trunk diameter up to 1.5 m (rarely larger); bark dark gray, furrowed; twigs slender, flexuous, glabrous or puberulent, the buds to 6 mm long, ovoid, acute, light brown, glabrous; leaf scars crescent-shaped, slightly elevated, with 3 bundle scars; leaves obovate to oval-oblong, acuminate at the apex, the asymmetrical base cuneate on one side, rounded on the other, doubly serrate, glabrous or sparsely scabrous on the upper surface, glabrous to soft-pubescent on the lower surface, to 15 cm long, up to half as broad, becoming bright yellow in the autumn, the petioles to 6 mm long, glabrous or nearly so, the caducous stipules glabrous or pubescent; flowers fascicled, the pedicels slender, glabrous or puberulent, 1.2–2.5 cm long, jointed near the base; calyx campanulate, to 4 mm long, 5- to 9-lobed, the lobes oblong, obtuse, green, puberulent; stamens 5–9, exserted, the filaments stramineous to light brown, the anthers bright red; samaras ovoid to oblongoid, to 1.2 cm long, subtended by the persistent calyx, the apex deeply notched, the base stipitate, the wings reticulate-nerved, ciliate on the margins.

COMMON NAMES: American Elm; White Elm.

HABITAT: Rich woods, wooded floodplains.

RANGE: Newfoundland to Saskatchewan, south to east-central Texas and Florida.

ILLINOIS DISTRIBUTION: Common throughout the state; in every county.

The American elm is a popular shade tree because of its beautiful widely spreading crown. It is becoming

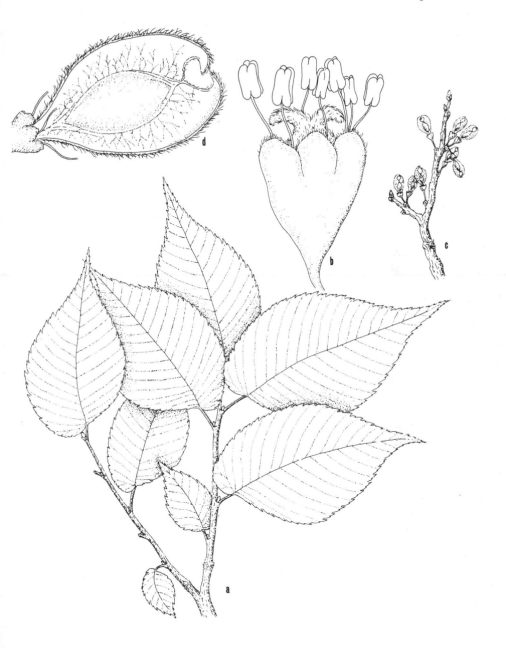

29. *Ulmus americana* (American Elm). *a*. Leafy branch, ×1. *b*. Flower, ×5. *c*. Twig, with fruits, ×½. *d*. Fruit, ×6.

uncommon in Illinois primarily because of the Dutch elm disease.

The leaves of this species are smaller, smoother, and shinier than those of the slippery elm. The similar-appearing Siberian elm has singly toothed leaves. A piece of outer bark of the American elm, when broken to show a cross-section, has alternating layers of light and dark gray.

The wood is heavy, hard, and tough, and is used for making boxes, barrels, and furniture.

The flowers bloom from February to April, while the fruits mature as the leaves begin to unfold.

3. **Ulmus pumila** L. Sp. Pl. 226. 1753. *Fig. 30.*

Tree to 25 m tall, usually much smaller, the crown rounded, the trunk diameter up to 1 m; bark gray to brown, with rough ridges separated by shallow fissures; twigs slender, gray or brown, glabrous or nearly so, the buds to 6 mm long, rounded, glabrous, dark reddish-brown; leaf scars crescent-shaped, slightly elevated, with 3 bundle scars; leaves oblong-lanceolate to elliptic, short-acuminate at the apex, rounded or cuneate at the slightly asymmetrical base, singly serrate, glabrous or nearly so on both surfaces, or pubescent in the axils of the veins beneath, to 8 cm long, to 4 cm broad, the petioles less than 1 cm long, glabrous, the stipules glabrous or nearly so; flowers fascicled, the pedicels slender and very short, glabrous; calyx campanulate, to 4 mm long, 5- to 9-lobed; stamens 4–5, exserted; samaras obovoid, to 1.2 cm long, usually a little broader, subtended by the persistent calyx, the apex notched, the wings reticulate-nerved, glabrous.

COMMON NAME: Siberian Elm.

HABITAT: Disturbed areas.

RANGE: Native of Asia; naturalized primarily in the central United States.

ILLINOIS DISTRIBUTION: Common to infrequent in most parts of the state.

The Siberian elm probably has been planted in nearly every Illinois community. It is particularly popular because of its resistance to disease. Seedlings come up in vacant lots, against buildings, along fences, and other places where a lawnmower will not reach.

In the west-central United States, this species is planted for a windbreak.

This species is frequently confused with the less common and

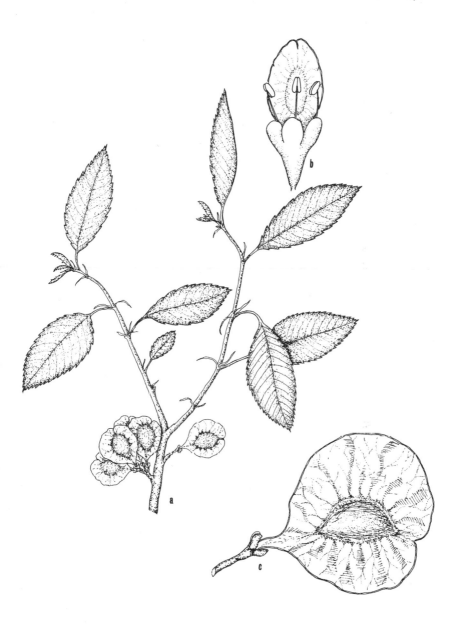

30. Ulmus pumila (Siberian Elm). *a.* Leafy branch, with fruits, ×1. *b.* Flower, with emerging fruit, ×5. *c.* Fruit, ×5.

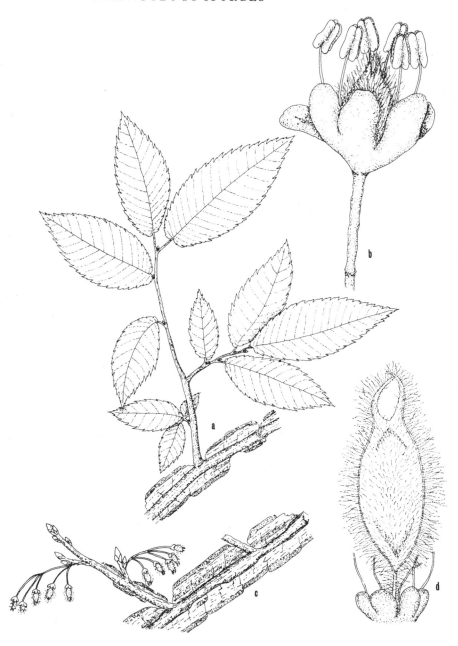

31. *Ulmus alata* (Winged Elm). *a*. Leafy branch, × 1. *b*. Flower, × 5. *c*. Twig, with fruits, × ½. *d*. Fruit, × 7½.

rarely escaped Chinese elm (*U. parvifolia* Jacq.). The Siberian elm flowers during May, while the Chinese elm blooms in August and September.

4. **Ulmus alata** Michx. Fl. Bor. Am. 1:173. 1803. *Fig. 31.*
Tree to 20 m tall, usually much smaller, the crown mostly oblong, the trunk diameter up to 0.7 m; bark light reddish-brown, with flat ridges separated by shallow fissures; twigs slender, light brown to gray, glabrous or nearly so, usually winged with two thin, corky wings, the buds ovoid, acute, dark brown, glabrous or puberulent, to 4 mm long; leaf scars crescent-shaped, slightly elevated, with 3 bundle scars; leaves oblong-lanceolate to oblong-ovate, acute or acuminate at the apex, cuneate or rounded at the slightly asymmetrical base, doubly serrate, pubescent at first but becoming glabrous or nearly so at maturity on the upper surface, softly pubescent below, even at maturity, to 9 cm long, to 4 cm broad, the petioles pubescent, to 8 mm long, the stipules narrow; flowers few in a fascicle, the pedicels slender and pendulous; calyx campanulate, to 3 mm long, 5- to 9-lobed, the lobes ovate, glabrous; stamens 5–9, exserted; samaras oblong, to 8 mm long, to 6 mm broad, subtended by the persistent calyx, the apex bifid, the base stipitate, the wings long-ciliate.

COMMON NAMES: Winged Elm; Wahoo.
HABITAT: Rocky upland woods, bluff-tops, less commonly in low ravines.
RANGE: Virginia to central Missouri, south to Texas and Florida.
ILLINOIS DISTRIBUTION: Occasional to common in the southern one-fourth of the state, apparently absent elsewhere.
The winged elm is a conspicuous species with its broad, corky wings. In specimens where the wings are greatly developed, the wings may measure as much as 2.5 cm across. A few specimens observed may be nearly destitute of wings.

Winged elm is one of the characteristic trees of exposed sandstone bluff-tops along the Shawneetown Ridge across southern Illinois, where it is nearly always associated with *Quercus marilandica, Q. stellata, Juniperus virginiana,* and *Vaccinium arboreum.* Under such xeric conditions, the winged elm is a tree of small, gnarled stature.

Occasionally winged elm grows in wooded ravines, where it is associated with *Celtis occidentalis* and *Gleditsia triacanthos*. In the ravines, the winged elm may reach a height of twenty meters.

At a few stations in southwestern Illinois, winged elm may occur in pin oak flatwoods.

The leaves turn a dull yellow in the autumn.

The wood, which is hard and heavy but difficult to split, is used primarily in the making of tool handles.

The flowers appear in February and March, while the fruits mature as the leaves begin to unfold.

5. **Ulmus thomasii** Sarg. Silva 14:102. 1902. *Fig. 32.*

Ulmus racemosa Thomas, Am. Journ. Sci. 19:170. 1831, non Borkh. (1800).

Tree to 25 m (in Illinois), the crown narrowly oblong, the trunk diameter up to 0.8 m; bark gray, with flat ridges separated by deep fissures; twigs slender, brown or gray, glabrous or puberulent, sometimes with 2–4 irregular, corky wings, particularly with age, the buds ovoid, acute, brown, pilose, and ciliate, to 6 mm long; leaf scars crescent-shaped, slightly elevated, with 3 bundle scars; leaves oval to obovate, abruptly acuminate at the apex, cuneate or rounded at the slightly asymmetrical base, doubly serrate, glabrous and lustrous on the upper surface, softly pubescent beneath, to 15 cm long, to 10 cm broad, turning bright yellow in the autumn, the petioles to 1 cm long, pubescent, the stipules ovate-lanceolate; flowers few in cymose racemes, the pedicels slender and pendulous; calyx campanulate, to 3 mm long, 5- to 8-lobed, the lobes oblong, obtuse; stamens 5–8, exserted, the anthers purple; samaras oval to ovoid, to 1.5 cm long and broad, subtended by the persistent calyx, the apex very shallowly notched, the wings obscurely nerved, pubescent, ciliate.

COMMON NAMES: Rock Elm: Cork Elm.

HABITAT: Rich woods, near streams.

RANGE: Quebec to South Dakota, south to eastern Kansas and Tennessee.

ILLINOIS DISTRIBUTION: Uncommon in the northern half of the state; absent elsewhere.

In general, the corky branches of the rock elm are not as intricately beautiful as those of the winged elm because of their irregular occurrence and growth. Usually only the oldest branchlets develop the corky outgrowths.

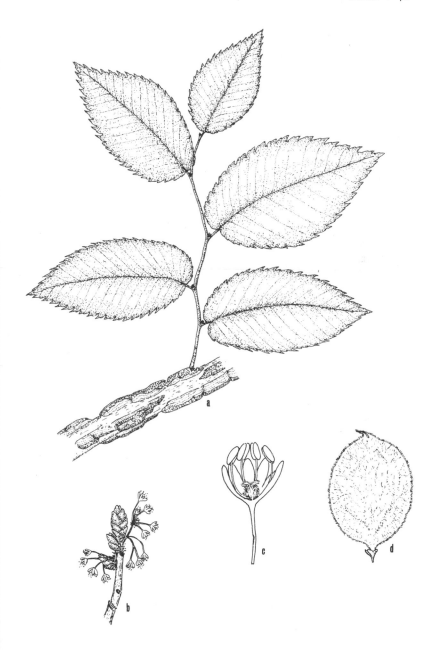

32. *Ulmus thomasii* (Rock Elm). *a*. Leafy branch, ×1. *b*. Twig, with bud or flower, ×1. *c*. Flower, with front stamens and front sepals removed, ×5. *d*. Fruit, ×5.

The leaves of the rock elm are similar to but smaller than those of the American elm.

The strong, tough, flexible wood makes this species valuable in the manufacture of wooden farm implements and tool handles.

The flowers bloom from March to May, while the fruits mature when the leaves are about half grown.

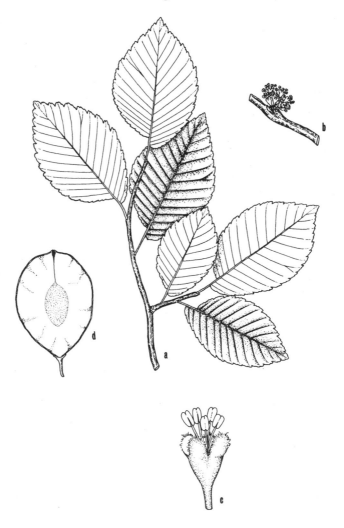

33. *Ulmus procera* (English Elm). *a.* Leafy branch, × ¼. *b.* Cluster of flowers, × 1. *c.* Flower, × 5. *d.* Fruit, × 2.

6. **Ulmus procera** Salisb. Prodr. 391. 1796. *Fig. 33.*

Tree to 15 m (in Illinois), the crown irregularly rounded, the trunk diameter up to 0.3 m (in Illinois); bark gray to brown, with flat ridges separated by deep fissures; twigs slender, brown or gray, pubescent at least when young, occasionally corky-winged, the buds ovoid, subacute, dark reddish-brown, more or less pubescent, to 6 mm long; leaf scars crescent-shaped, slightly elevated, with 3 bundle scars; leaves elliptic to lance-ovate, abruptly short-acuminate at the apex, rounded or cuneate at the asymmetrical base, doubly serrate, scabrous on the upper surface, pubescent at least in the axils of the veins beneath, to 8 cm long, to 4 cm broad, the petioles to 5 mm long; flowers in fascicles, on short pedicels; calyx campanulate, to 3.5 mm long, 3- to 5-lobed; stamens 3–5, exserted; samaras orbicular, to 1.2 cm in diameter, subtended by the persistent calyx, the apex with a short, closed notch, the wings glabrous.

COMMON NAME: English Elm.

HABITAT: Along road near abandoned farm.

RANGE: Native of Europe; occasionally planted and sometimes spreading by means of suckers.

ILLINOIS DISTRIBUTION: Jackson Co.: along blacktop road between Carbondale and Giant City State Park, June 26, 1970, *R. Mohlenbrock 19256.*

The English elm is sometimes planted as an ornamental in Illinois. It has the ability to produce an abundance of suckers. The specimen cited above was made from a colony of several specimens. Most of the plants in the colony had some branchlets with corky ridges.

The flowers bloom during April.

2. *Planera J. F. Gmel.*–Water Elm

Tree with scaly bark; twigs unwinged, with small, solid pith; bud scales 2-ranked; leaves alternate, simple, serrate, the stipules caducous; flowers monoecious or polygamous, fascicled; calyx campanulate, 4- to 5-lobed; petals absent; stamens few, free, exserted beyond the calyx; pistil 1, the ovary superior, stipitate, 1-locular, 1-ovulate, the styles deeply 2-lobed; fruit a 1-seeded, nutlike drupe.

Only the following species comprises the genus.

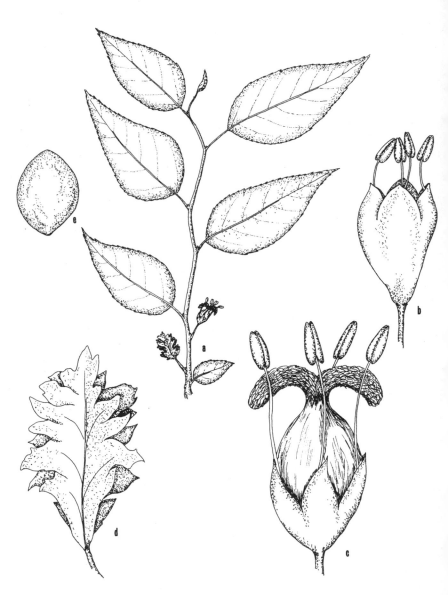

34. Planera aquatica (Water Elm). *a.* Leafy branch, with flowers, ×1. *b.* Staminate flower, ×5. *c.* Perfect flower, ×5. *d.* Fruit, ×5. *e.* Seed, ×5.

1. **Planera aquatica** [Walt.] J. F. Gmel. Syst. Nat., ed. 13, 2:150. 1791. *Fig.* 34.

Anonymos aquatica Walt. Fl. Car. 230. 1788.

Planera gmelinii Michx. Fl. Bor. Am. 2:248. 1803.

Tree to 10 m tall, the crown broad and irregular, the trunk diameter up to 0.4 m; bark brown or gray, with large scales; twigs slender, puberulent, becoming scaly, brown or gray, the buds subglobose, acute, brown or reddish-brown, puberulent; leaf scars suborbicular, slightly elevated, with several bundle scars; leaves oblong-ovate, acute at the apex, rounded or subcordate at the slightly asymmetrical base, singly crenate-serrate, pubescent when young, becoming glabrous or nearly so and scabrellous at maturity, to 6 cm long, to 2.5 cm broad, the petioles to 6 mm long, puberulent, the stipules ovate, red, caducous; staminate flowers fasciculate from the outer bud scales on twigs of the previous year, on very short pedicels; perfect flowers in clusters of 1–3 from the axils of the current leaves, on elongated pedicels; calyx campanulate, deeply 4- to 5-lobed, greenish-yellow; stamens few, exserted; drupe oblongoid, stipitate, subtended by the persistent calyx, tipped by the persistent styles, pale brown, tuberculate, to 8 mm long, with one dark brown, ovoid seed.

COMMON NAMES: Water Elm; Planer Tree.

HABITAT: Swampy woods.

RANGE: North Carolina to southeastern Missouri, south to Texas and northern Florida.

ILLINOIS DISTRIBUTION: Known only from Alexander, Johnson, Massac, Pope, and Pulaski counties in the extreme southern tip of the state.

The small, pointed buds, scaly bark, and drupes distinguish the water elm from the genus *Ulmus* in Illinois. Beneath the scaly outer bark is a reddish-brown inner bark.

The tree is too rare, and the wood is too light and soft, to be of any commercial value.

The flowers are borne in March and April.

3. *Celtis* L.–Hackberry

Trees or shrubs, usually with warty bark; twigs with chambered pith; bud scales 2-ranked; leaves alternate, simple, serrate, the stipules caducous; flowers polygamous, solitary or fascicled; calyx deeply 4- to 5-lobed, or the sepals free; petals absent; stamens 5–

6, free, exserted; pistil one, the ovary superior, sessile, 1-locular, 1-ovulate, the styles deeply 2-lobed; fruit a 1-seeded drupe.

Celtis is a genus of about seventy species found primarily in the Northern Hemisphere.

Most species of the genus are attacked by a gall insect which causes the production of gnarled broomlike clusters of branchlets, a disfiguration known as witch's-broom.

The Illinois taxa of *Celtis* are extremely variable and seemingly intergrade in an almost hopeless manner. Usually exact identification can only be made if mature drupes are present.

KEY TO THE SPECIES OF Celtis IN ILLINOIS

1. Leaves conspicuously several-toothed _____ 2
1. Leaves entire or only sparingly toothed _____ 5
   2. Leaves harshly scabrous on the upper surface _ 1. *C. occidentalis*
   2. Leaves smooth or nearly so on the upper surface _____ 3
3. Most of the leaves more than half as long as broad, very strongly asymmetrical at the base _____ 1. *C. occidentalis*
3. Most of the leaves less than half as long as broad, only slightly asymmetrical at the base _____ 4
   4. Drupe 8–11 mm long, dark purple or dark brown _____
   _____ 1. *C. occidentalis*
   4. Drupe 5–7 mm long, orange, pale brown, or red _ 2. *C. laevigata*
5. Leaves harshly scabrous on the upper surface _____ 6
5. Leaves smooth or nearly so on the upper surface _____ 7
   6. Leaves mostly over half as broad as long _____ 3. *C. tenuifolia*
   6. Leaves mostly less than half as broad as long ____ 2. *C. laevigata*
7. Leaves more than half as broad as long, acute to short-acuminate at the tip _____ 3. *C. tenuifolia*
7. Leaves less than half as broad as long, long-tapering at the apex _____
_____ 2. *C. laevigata*

1. Celtis occidentalis L. Sp. Pl. 1044. 1753.

Trees to 30 m tall, the crown broad and irregular, the trunk diameter up to 1.5 m; bark dark brown or gray, smooth except for numerous wartlike outgrowths; twigs slender, gray, glabrous, the buds slenderly ovoid, acute, brown, pubescent, appressed against the twigs, to 6 mm long; leaf scars crescent-shaped, slightly elevated, with 3 bundle scars; leaves ovate to oblong-ovate, acuminate at the apex, cuneate to truncate to rounded at the asymmetrical base, serrate, smooth or harshly scabrous on the upper surface, glabrous or pilose along the veins on the lower surface, to 12 cm long, to 9 cm broad, coriaceous to membranous, the petiole glabrous or pubescent, to 1.5 cm long; staminate flowers several in axillary fascicles; pistillate flowers 1–3 in the leaf axils; sepals 4–5, free or nearly so, linear-oblong, greenish-yellow, not persistent in fruit; drupes globose to ovoid, purple, black, or orange-red at maturity, 8–11 mm long, the seed 7–9 mm long.

Three varieties may usually be distinguished in Illinois.

1. Leaves harshly scabrous on the upper surface; drupes orange-red at maturity _____ 1a. *C. occidentalis* var. *occidentalis*
1. Leaves more or less smooth on the upper surface; drupes purple or black at maturity. _____
    2. Most or all the leaves less than half as broad as long _____ _____ 1b. *C. occidentalis* var. *pumila*
    2. Most or all the leaves less than half as broad as long _____ _____ 1c. *C. occidentalis* var. *canina*

1a. Celtis occidentalis L. var. **occidentalis** *Fig. 35a–d.*

Celtis crassifolia Lam. Encycl. 4:138. 1796.
Celtis occidentalis L. var. *crassifolia* (Lam.) Gray, Man. Bot. ed. 2, 297. 1856.

Leaves harshly scabrous on the upper surface; drupes orange-red at maturity.

COMMON NAME: Hackberry.
HABITAT: Low woods.
RANGE: Massachusetts to Idaho, south to Oklahoma and northern Florida.
ILLINOIS DISTRIBUTION: Common throughout the state. In addition to the characters given in the key, a few others should be mentioned for var. *occidentalis*. The leaves of this variety tend to be coriaceous. The pedicels of the drupe never exceed 1.5 cm in length.

The heavy wood is sometimes used for fence posts and as a source of fuel.

The flowers appear in April and May. The fruit ripens in September and October.

1b. Celtis occidentalis L. var. **pumila** (Pursh) Gray, Man. Bot. ed. 2, 397. 1856. *Fig. 35e.*

Celtis pumila Pursh, Fl. Am. Sept. 200. 1814.

35. *Celtis occidentalis* (Hackberry). *a*. Leafy branch, with fruits, × 1. *b*. Leaf variation, × 1. *c*. Flower, × 5. *d*. Fruit, × 2. var. *pumila* (Small Hackberry). *e*. Leaf, × 1. var. *canina* (Hackberry). *f*. Leaf, × 1.

Leaves more or less smooth on the upper surface, most or all of them more than half as broad as long; drupes purple or black.

COMMON NAME: Small Hackberry.

HABITAT: Low woods; bluffs.

RANGE: Quebec to North Dakota, south to Oklahoma and Georgia.

ILLINOIS DISTRIBUTION: Occasional in most parts of the state.

This is the least common of the three varieties of *C. occidentalis* in Illinois. Its smaller stature makes it easily confused with *C. tenuifolia*, but this latter species has the leaves mostly entire. *Celtis laevigata* var. *smallii*, which has toothed leaves, has smaller fruits and seeds than *C. occidentalis* var. *pumila*.

The flowers appear in April and May, while the drupes mature in September and October.

1c. **Celtis occidentalis** L. var. **canina** (Raf.) Sarg. Bot. Gaz. 67:217. 1919. *Fig.* 35*f*.

Celtis canina Raf. Am. Monthly Mag. 2:43. 1817.

Leaves more or less smooth on the upper surface, most or all of them less than half as broad as long; drupes purple or black.

COMMON NAME: Hackberry.

HABITAT: Low woods.

RANGE: Quebec to Utah, south to Oklahoma and Georgia.

ILLINOIS DISTRIBUTION: Common throughout the state.

This seems to be the most common of the three varieties of *C. occidentalis* in Illinois.

The leaves of var. *canina* are more symmetrical at the base than those of any other varieties of *C. occidentalis* in Illinois.

The flowers are borne in April and May, the drupes in September and October.

2. **Celtis laevigata** Willd. Berl. Baumz., ed. 2, 81. 1811.

Trees to 30 m tall, the oblong crown rather open, the trunk diameter up to 0.75 m; bark gray, smooth except for numerous wartlike outgrowths; twigs slender, gray, glabrous, the buds ovoid, acute, brown, glabrous or nearly so, appressed against the twigs, to 3 mm long; leaf scars crescent-shaped, slightly elevated, with 3 bundle scars; leaves lanceolate to lance-oblong to ovate, acute to acuminate at the apex, cuneate to rounded to subcordate at the usually asymmetrical base, entire or sparsely toothed or regularly toothed, smooth or scabrous on the upper surface, glabrous or pubescent at least along the veins on the lower surface, to 10 cm long, to 4 cm broad, coriaceous to membranous, the petioles to 1 cm long, glabrous or pubescent; staminate flowers several in axillary fascicles; pistillate flowers 1–3 in the leaf axils; sepals 4–5, free or nearly so, ovate-lanceolate, greenish-yellow, not persistent in fruit; drupes ellipsoid to globose, orange, red, or brownish at maturity, 5–8 mm long, the seed 4–7 mm long.

Three varieties may usually be differentiated in Illinois.

1. Leaves more or less smooth on the upper surface; petioles essentially glabrous _____ 2
1. Leaves harshly scabrous on the upper surface; petioles pubescent _____ 2c. *C. laevigata* var. *texana*
    2. Leaves entire or sparingly toothed _____
    _____ 2a. *C. laevigata* var. *laevigata*
    2. Leaves regularly toothed _____ 2b. *C. laevigata* var. *smallii*

2a. Celtis laevigata Willd. var. **laevigata** *Fig. 36a–c.*

Celtis mississippiensis Bose, Encycl. Met. Agr. 7:577. 1822.

Celtis occidentalis L. var. *mississippiensis* (Bosc) Schneck, Ann. Rep. Geol. Surv. Ind. 7:559. 1876.

Leaves more or less smooth on the upper surface, entire or sparingly toothed; petioles essentially glabrous.

COMMON NAME: Sugarberry.

HABITAT: Low woods.

RANGE: Virginia to Oklahoma, south to Texas and Florida; Mexico.

ILLINOIS DISTRIBUTION: Occasional to common in the southern half of the state; also Peoria County.

The sugarberry is a species of low woodlands. The typical variety is more common than the other two varieties of *C. laevigata* in Illinois. It generally can be recognized by its nearly entire, membranous leaves.

The flowers are formed in April and May. The drupes mature in September and October.

36. *Celtis laevigata* (Sugarberry). *a.* Leafy twig, ×1. *b.* Leaf variation, ×1. *c.* Flower, ×5. var. *smallii* (Toothed Sugarberry). *d.* Leaf and fruit, ×1. var. *texana* (Cliff Sugarberry). *e.* Leaf, ×1.

2b. Celtis laevigata Willd. var. **smallii** (Beadle) Sarg. Bot. Gaz. 67:223. 1919. *Fig. 36d.*

Celtis smallii Beadle in Small, Fl. S.E.U.S. 365. 1903.

Leaves more or less smooth on the upper surface, regularly toothed; petioles essentially glabrous.

COMMON NAME: Toothed Sugarberry.
HABITAT: Low woods.
RANGE: Virginia to Missouri, south to Louisiana and Florida.
ILLINOIS DISTRIBUTION: Occasional in the extreme southern counties of the state.
This variety has the toothed leaves of *C. occidentalis* but the small drupes and seeds of *C. laevigata*. The leaves also tend to be narrower than those of *C. occidentalis*.

The flowers are borne in April and May, the drupes in September and October.

2c. **Celtis laevigata** Willd. var. **texana** (Scheele) Sarg. Bot. Gaz. 67:223. 1919. *Fig. 36e.*

Celtis texana Scheele, Linnaea 22:146. 1849.

Leaves harshly scabrous on the upper surface, essentially entire; petioles pubescent.

COMMON NAME: Cliff Sugarberry.
HABITAT: Dry woods and cliffs.
RANGE: Southern Illinois to Kansas, south to New Mexico and Texas.
ILLINOIS DISTRIBUTION: Known only from the southern one-fifth of the state.
This variety is more reminiscent of *C. tenuifolia* var. *georgiana* because of its bluff habitat, its coriaceous leaves, and its pubescent petioles. It is kept as a variety of *C. laevigata* primarily on the basis of the slender, elongated leaves.

The flowers appear in April and May, while the fruits are matured in September and October.

3. **Celtis tenuifolia** Nutt. Gen. Am. 1:202. 1818.

Small trees to 7 m tall (in Illinois), the crown irregular, the trunk diameter up to 0.3 m; bark gray to reddish-brown, smooth except for some warty outgrowths; twigs slender, reddish-brown, glabrous or pubescent, the buds ovoid, acute, reddish-brown, pubescent, to

2 mm long; leaf scars crescent-shaped, slightly elevated, with 3 bundle scars; leaves ovate, acute to short-acuminate at the apex, rounded or cordate at the more or less asymmetrical base, entire or sparsely toothed, smooth or scabrous on the upper surface, more or less pubescent beneath, coriaceous or membranous, the petioles to 6 mm long, pubescent; staminate flowers several in axillary fascicles; pistillate flowers 1–3 in the leaf axils; sepals 4–5, free or nearly so, lanceolate, greenish-yellow, not persistent in fruit; drupes globose, orange, red, or brown at maturity, 5–8 mm long, the seed 5–7 mm long.

Two varieties may be differentiated in Illinois.

1. Leaves more or less smooth on the upper surface, membranous, the petioles glabrous or nearly so _ _ _ _ _ _ 3a. *C. tenuifolia* var. *tenuifolia*
1. Leaves harshly scabrous on the upper surface, coriaceous, the petioles densely pubescent _ _ _ _ _ _ _ _ _ _ _ _ 3b. *C. tenuifolia* var. *georgiana*

3a. Celtis tenuifolia Nutt. var. **tenuifolia** *Fig. 37a–e.*

Celtis pumila Pursh var. *deamii* Sarg. Bot. Gaz. 67:227. 1919.

Leaves more or less smooth on the upper surface, membranous, the petioles glabrous or nearly so.

COMMON NAME: Dwarf Hackberry.

HABITAT: Dry woods and cliffs.

RANGE: Pennsylvania to Missouri, south to Louisiana and Florida.

ILLINOIS DISTRIBUTION: Confined mostly to the southern half of the state.

This is the plant which early workers had called *C. pumila* Pursh, but Pursh's epithet is assignable to *C. occidentalis*.

This variety of the dwarf hackberry is less common than var. *georgiana*. Both occur on dry cliffs.

The flowers appear in April and May, while the fruits are ripe in September and October.

3b. Celtis tenuifolia Nutt. var. **georgiana** (Small) Fern. & Schub. Rhodora 50:160. 1948. *Fig. 37f.*

Celtis georgiana Small, Bull. Torrey Club 24:439. 1897.

Leaves harshly scabrous on the upper surface, coriaceous, the petioles densely pubescent.

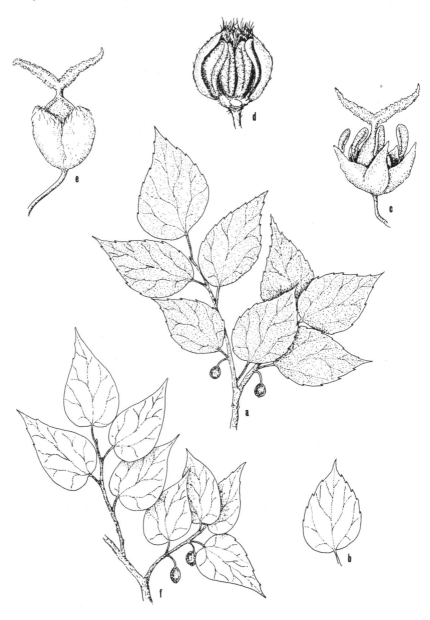

37. *Celtis tenuifolia* (Dwarf Hackberry). *a.* Leafy branch, with fruits, × ½. *b.* Leaf with teeth, × ½. *c.* Perfect flower, × 5. *d.* Staminate flower, with one sepal and one anther removed, × 5. *e.* Pistillate flower, × 5. var. *georgiana.* *f.* Leafy branch, with fruits, × ½.

COMMON NAME: Dwarf Hackberry.
HABITAT: Dry woods and cliffs.
RANGE: Virginia to Kansas, south to Oklahoma, Louisiana, and Georgia.
ILLINOIS DISTRIBUTION: Mostly in the southern one-fifth of the state.
This is the common variety of *C. tenuifolia* in Illinois, but it is in no sense of the word a common plant.
The flowers are borne in April and May. The fruits are produced in September and October.

MORACEAE–MULBERRY FAMILY

Trees, shrubs, vines, or herbs, often with latex; leaves alternate or opposite, simple, with 2 stipules; flowers unisexual, actinomorphic, variously arranged; sepals usually 4 or 5, free or united at the base; petals absent; stamens 1, 2, 4, or 5, free; pistil 1, the ovary superior or inferior, 2-carpellate, 1-locular, 1-ovulate, the styles 1–2; fruit a drupe or cluster of drupes or an achene.

As considered in this work, the Moraceae include the Cannabinaceae. Some authors prefer to treat the Cannabinaceae, including *Cannabis* and *Humulus*, as a separate family on the basis of the herbaceous habit and presence of achenes.

In all the World, there are about seventy-five genera and more than one thousand species in the Moraceae. Most of these live in the tropics.

The family has much economic importance. *Ficus*, the largest genus, produces figs, is the source of various grades of rubber, and is grown as an ornamental. The tropical breadfruit and jackfruit belong to the genus *Artocarpus*. *Morus*, the mulberry, in addition to its edible fruits, is important in the silk industry. *Humulus* is the source of hops, while *Cannabis* not only is marijuana but is also an important hemp-producing plant used for cordage. Many other tropical genera have edible fruits.

KEY TO THE GENERA OF Moraceae IN ILLINOIS

1. Trees or shrubs; latex present _____ 2
1. Herbs or vines; latex absent _____ 4
    2. Leaves toothed, often lobed; branches without spines; fruit 1–2 cm in diameter, white, orange, or red _____ 3
    2. Leaves entire, unlobed; branches with short spines; fruit 10–20 cm in diameter, greenish-yellow _____ 3. *Maclura*

3. Leaves glabrous or short-hairy on the lower surface, not velvety; most or all the petioles less than 3 cm long, glabrous or appressed-pubescent; bark roughened _____ 1. *Morus*
3. Leaves velvety on the lower surface; most or all the petioles more than 3 cm long, pilose; bark smooth _____ 2. *Broussonetia*
4. Stems climbing or trailing; leaves simple, although often lobed _____ 4. *Humulus*
4. Stems erect; leaves compound _____ 5. *Cannabis*

1. *Morus* L.–Mulberry

Monoecious or dioecious trees or shrubs with usually scanty latex, the pith continuous, the bud scales 2-ranked; leaves alternate, sometimes lobed; stipules 2, enclosing the bud and leaving a circular scar upon falling; flowers unisexual, arranged in axillary spikes; sepals 4, united at the base; petals absent; stamens 4, free; ovary superior, 1-locular, with 1 pendulous ovule; styles 2; drupes clustered together among the persistent fleshy perianth, forming a compound fruit.

The genus is composed of about ten very variable species native to the Northern Hemisphere.

KEY TO THE SPECIES OF Morus IN ILLINOIS

1. Lower surface of leaves pubescent on the blade as well as on the veins _____ 1. *M. rubra*
1. Lower surface of leaves glabrous, or pubescent only on the veins, or the hairs confined to axillary tufts _____ 2. *M. alba*

1. Morus rubra L. Sp. Pl. 986. 1753. *Fig. 38.*

Trees to 20 m tall, the crown broadly rounded, the trunk diameter up to 1 m; bark dark brown, divided into long, scaly plates; twigs slender, sometimes flexuous, orange-brown to dark brown, glabrous or pubescent; buds ovoid, acute to obtuse, brown, to 6 mm long, the scales puberulent and ciliate; leaf scars circular, slightly elevated, with several bundle scars; leaves ovate, acute to short-acuminate at the apex, truncate or cordate at the base, toothed, sometimes 2-lobed, to 10 cm long, to 8 cm broad, scabrous on the upper surface, pubescent on the lower surface, the petioles glabrous or tomentose, to 3 cm long; flowers usually dioecious, the staminate in narrow spikes to 5 cm long, the pistillate in oblongoid spikes to 2 cm long; sepals 4, nearly free, the lobes oblong, obtuse, pubescent; stamens 4, free; ovary superior, glabrous; fruit com-

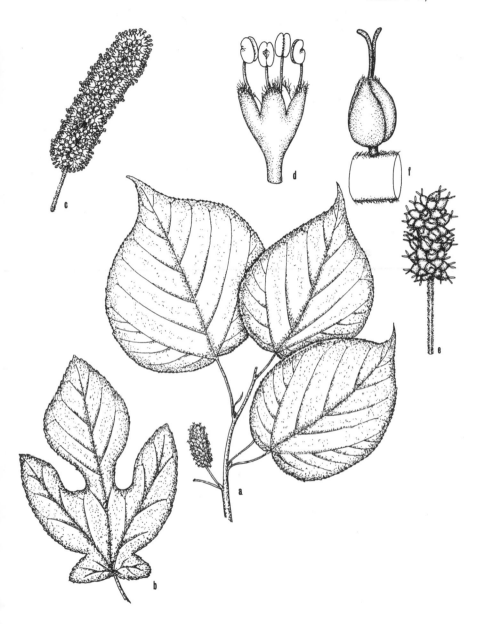

38. Morus rubra (Red Mulberry). *a.* Branch, with leaves and fruit, ×¼. *b.* Leaf, ×¼. *c.* Cluster of staminate flowers, ×2. *d.* Staminate flower, ×7½. *e.* Cluster of pistillate flowers, ×2. *f.* Pistillate flower, ×12½.

pound, to 5 cm long, dark purple, sweet, juicy, composed of numerous drupes, each drupe 1-seeded.

COMMON NAME: Red Mulberry.
HABITAT: Woodlands, fields, roadsides.
RANGE: Vermont to South Dakota, south to Texas and Florida.
ILLINOIS DISTRIBUTION: Common throughout the state. Red mulberry is best distinguished from white mulberry by the pubescence on the lower surface of the leaves. The upper surface of the red mulberry leaves is usually rough to the touch. The color of the fruit is not necessarily distinctive since *M. alba* var. *tatarica* also has dark purple fruits. In *M. alba* var. *tatarica*, however, the fruit rarely exceeds 1 cm in length.

Leaf shape is extremely variable in this species. Some leaves are unlobed, others may be two-lobed and mitten-shaped, while others may be three-lobed. Size of leaves varies considerably, as well.

Since mulberries are sweet and juicy, they are sought after by wildlife and even made into jams, jellies, and pies by humans.

The heavy and durable wood, which is slightly soft, is useful in making barrels and fence posts.

The flowers of *M. rubra* bloom in April and May. The fruits ripen in June and July.

The following chart compares the distinguishing features of the red mulberry and the white mulberry.

Comparison of Features

Red Mulberry	White Mulberry
Leaves dull, large, 3 to 10 inches long, always finely hairy below between the veins.	Leaves shiny, smaller, 2 to 4 inches long, never hairy below between the veins, although usually hairy on the veins below.
Leaves almost always hairy above.	Leaves almost always absolutely smooth above.
Lateral lobes of leaves often pointed.	Lateral lobes of leaves always blunt.
Buds greenish-brown; scales with darker brown borders.	Buds reddish-brown.
Buds sometimes diverging from the twig.	Buds always close against the twig.
Fruit always red or purple.	Fruit usually red or purple, sometimes white.

Bark dark brown.   Bark light brown with orange
                                    tinge.

2. Morus alba L. Sp. Pl. 986. 1753.

Trees to 15 m tall, the crown broadly rounded, the trunk diameter
up to 0.8 m; bark light gray, covered with scaly plates; buds ovoid,
acute to obtuse, brown, to 6 mm long, the scales puberulent and
ciliate; leaf scars circular, slightly elevated, with several bundle
scars; leaves ovate, acute to short-acuminate at the apex, truncate
or cordate at the base, toothed, sometimes 3- to 7-lobed, to 20 cm
long, usually smaller, smooth or nearly so on the upper surface,
glabrous on the lower surface or pubescent on the veins or in the
axils of the veins, the petioles usually glabrous, to 4 cm long; flowers usually dioecious, the staminate in narrow spikes to 4.5 cm long,
the pistillate in oblongoid spikes to 3 cm long; sepals 4, nearly free,
the lobes oblong, obtuse, pubescent; stamens 4, free; ovary superior, usually glabrous; fruit compound, to 3 cm long, white to pink
to red to purple, sweet, juicy, composed of numerous drupes, each
drupe 1-seeded.

Two varieties are worthy of recognition in Illinois.

1. Fruits white to pink, usually over 1 cm long __ 2a. *M. alba* var. *alba*
1. Fruits red to purple, up to 1 cm long ____ 2b. *M. alba* var. *tatarica*

2a. Morus alba L. var. **alba** *Fig. 39.*
Morus alba L. var. *skeletoniana* Schneider, Handb. Laubholzk.
1:217. 1904.
Morus alba L. f. *skeletoniana* (Schneider) Rehder, Bibl. Cult.
Trees & Shrubs 147. 1949.
Fruits white to pink, usually over 1 cm long.

COMMON NAME: White Mulberry.
HABITAT: Fields, fencerows, roadsides, disturbed woods.
RANGE: Native of eastern Asia; naturalized throughout
most of the United States.
ILLINOIS DISTRIBUTION: Common throughout the state.
The white- or pink-fruited variety of *Morus alba* is far
more common in Illinois than the red- or purple-fruited
variety.
The fruits are sweet and edible.
Considerable variation in leaf-cutting is exhibited in Il-

39. *Morus alba* (White Mulberry). *a.* Leafy branch, with fruits, ×½. *b,c,d.* Leaf variations, ×½. *e.* Staminate flower, ×5. *f.* Pistillate flower, ×5.

linois. Specimens with several deeply divided lobes may be designated as f. *skeletoniana*.

The caterpillars of the silkworm moth feed upon the leaves of the white mulberry.

The weeping mulberry, with pendulous branches, is a variety (var. *pendula* Dipp.) of *Morus alba*. It is sometimes grown as an ornamental.

The white mulberry flowers in April and May. The fruits mature in June and July.

2b. Morus alba L. var. **tatarica** (L.) Loudon, Arb. & Fruct. 3:1368. 1838. Not illustrated.

Morus tatarica L. Sp. Pl. 986. 1753.

Fruits red to purple, up to 1 cm long.

COMMON NAME: Russian Mulberry.

HABITAT: Along fences and roads.

RANGE: Native of Europe and Asia; adventive particularly in the central United States.

ILLINOIS DISTRIBUTION: Occasional throughout the state.

In addition to the differences in fruit color and size, *M. alba* var. *tatarica* tends to be more of a shrub form.

2. *Broussonetia* L'Her.–*Paper Mulberry*

Dioecious trees or shrubs with latex, the pith continuous, except for a thin green diaphragm at each node, the bud scales 2-ranked; leaves alternate, rarely opposite, often lobed; flowers unisexual, the staminate arranged in pendulous spikes, the pistillate in headlike clusters; sepals 4, nearly free in the staminate flowers, forming a short tube in the pistillate flowers; petals absent; stamens 4, free; ovary superior, stipitate, 1-locular, with 1 ovule; style 2-cleft; drupes clustered into a globose fruit and protruding from it, with persistent sepals and styles.

Broussonetia is a genus of two Asiatic species.

Only the following species occurs in Illinois.

1. Broussonetia papyrifera (L.) L'Her, ex Vent. Tabl. Reg. Veg. 3:548. 1799. *Fig. 40.*

Morus papyrifera L. Sp. Pl. 986. 1753.

Papyrius papyrifera (L.) Kuntze, Rev. Gen. Pl. 629. 1891.

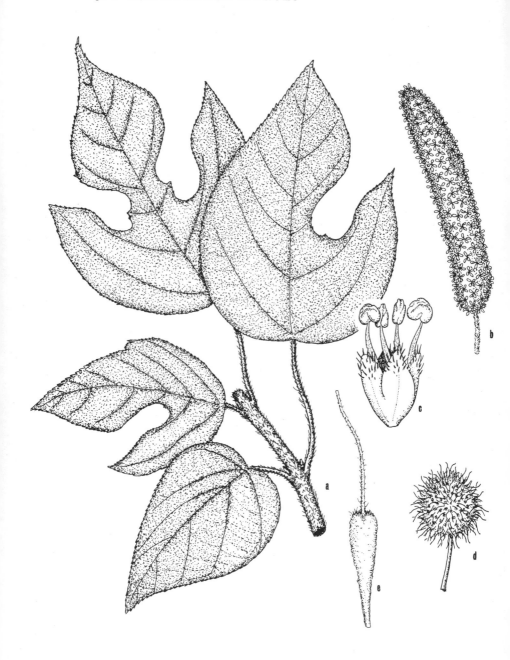

40. *Broussonetia papyrifera* (Paper Mulberry). *a.* Leafy branch, × ½. *b.* Staminate spike, × 1. *c.* Staminate flower, × 5. *d.* Pistillate head, × 1. *e.* Pistillate flower, × 5.

Tree to 15 m tall, the crown rounded and spreading, the trunk diameter to 0.3 m; bark gray with yellowish lines, more or less smooth; twigs slender, sometimes flexuous, gray, pubescent; buds ovoid, acute, brownish, to 6 mm long, the scales puberulent and ciliate; leaf scars narrow, slightly elevated, with 5 bundle scars; leaves ovate, acute at the apex, subcordate at the base, serrate, sometimes 2- to 3-lobed, to 20 cm long, scabrous on the upper surface, tomentose on the lower surface, the petioles pubescent, to 10 cm long; flowers dioecious; staminate flowers in spikes to 7 cm long, the sepals 4, nearly free, the stamens 4, free; pistillate flowers in heads up to 2 cm in diameter, the calyx 4-lobed, tubular, the ovary superior, pubescent, the styles long-exserted; fruit globose, compound, to 2 cm in diameter, the drupes orange or red, strongly protruding, tomentose.

COMMON NAME: Paper Mulberry.

HABITAT: Roadsides, disturbed woods.

RANGE: Native of Asia; naturalized in the eastern half of the United States.

ILLINOIS DISTRIBUTION: Known in the wild only from the southern one-fourth of the state.

The common name paper mulberry is derived from the use of the bark for paper-making in Asia. The pollen is reported to be a cause for hay fever.

3. Maclura Nutt.–Osage Orange

Dioecious tree with scanty latex, with axillary spines, the pith continuous, the bud scales more or less 2-ranked; leaves alternate, entire; flowers unisexual, the staminate in elongated racemes from short, spurlike branches, the pistillate in dense, globose, sessile heads from the leaf axils; calyx 4-parted; petals absent; stamens 4, free; ovary superior, sessile, 1-locular, with 1 ovule; style 1; drupes clustered into a large globose fruit, the persistent perianth fleshy.

Only the following species comprises the genus.

1. **Maclura pomifera** (Raf.) Schneid. Handb. Laubh. 1:806, 1906. *Fig. 41.*

Toxylon pomiferum Raf. Am. Monthly Mag. 2:118. 1817.

Maclura aurantiaca Nutt. Gen. N. Am. Pl. 2:234. 1818.

Tree to 15 m tall, the crown round-topped but irregular, the trunk diameter up to 0.4 m; bark light grayish-brown tinged with orange,

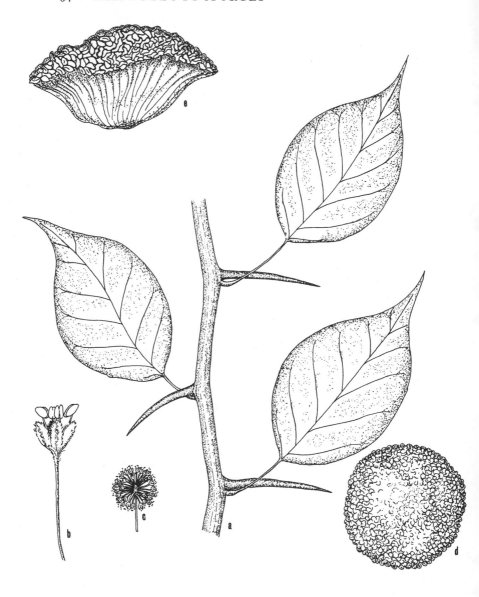

41. *Maclura pomifera* (Osage Orange). *a*. Leafy branch, × ½. *b*. Staminate flower, × 5. *c*. Pistillate head, × 1. *d*. Fruit, × ¼. *e*. Section of fruit, × ½.

breaking into shaggy strips; twigs slender to moderately stout, dull orange-brown, glabrous at maturity, usually with a spine at each leaf axil; buds globose, partly sunken in the twig, brown, ciliate; leaf scars half-round, slightly elevated, with 3 groups of bundle scars; leaves ovate to oblong-lanceolate, acuminate at the apex, subcordate, rounded, or cuneate at the base, entire, glabrous and lustrous at maturity, to 12 cm long, to 7 cm broad, the petioles pubescent, to 4 cm long; flowers dioecious; staminate flowers in racemes to 3 cm long, the calyx 4-lobed to the middle, pubescent, the stamens 4, free; pistillate flowers in globose heads to 2 cm in diameter, the calyx 4-lobed nearly to the base, pubescent, persistent; fruit a globose head to 12 cm in diameter, greenish-yellow, composed of numerous oblongoid drupes each notched at the apex and enclosed by the fleshy, persistent calyx, each drupe 1-seeded.

COMMON NAMES: Osage Orange; Hedge Apple.
HABITAT: Fencerows, roadsides, fields, disturbed woods.
RANGE: Native of the south-central United States; naturalized in the eastern half of the United States.
ILLINOIS DISTRIBUTION: Throughout the state.
Maclura pomifera is a species commonly planted along farm boundaries to serve the purpose of a fence.
The huge fruits are eaten by wild animals but have very little value to humans. It is said that they will ward off roaches and spiders, but I have no proof of that.
The strong, durable, heavy wood is employed in the making of railroad ties and fence posts. Indians of the Southwest, where this species is native, made bows from the wood.

The Osage orange flowers during May and June. The fruits are ripe in July and August.

4. *Humulus* L.–Hops

Dioecious twining herbs; leaves opposite, often palmately lobed, the stipules persistent; flowers unisexual, the staminate in axillary panicles, the pistillate in pendulous spikes; sepals 5, free, or united and entire; stamens 5, free; ovary superior, 1-locular, with 1 pendulous ovule; styles 2-cleft; fruits a cluster of achenes enclosed by inflated, persistent bracts.

Humulus is a genus of two species native to the Northern Hemisphere.

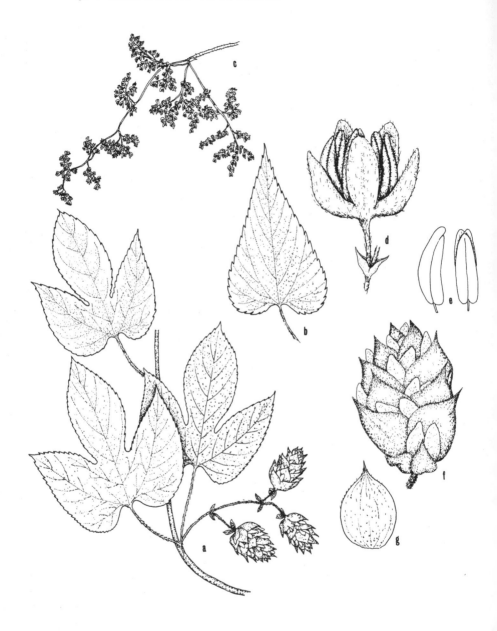

42. *Humulus lupulus* (Common Hops). *a.* Leafy branch, with fruits, ×½. *b.* Leaf variation, ×½. *c.* Staminate panicle, ×½. *d.* Staminate flower, ×5. *e.* Stamens, ×7½. *f.* Fruit, ×1½. *g.* Seed, ×3.

KEY TO THE SPECIES OF Humulus IN ILLINOIS

1. Most of the leaves 3-lobed, with resinous glands beneath _____
   _____ 1. *H. lupulus*
1. Most of the leaves 5- to 7-lobed, without resinous glands beneath
   _____ 2. *H. japonicus*

1. **Humulus lupulus** L. Sp. Pl. 1028. 1753. *Fig. 42.*

Humulus americanus Nutt. Journ. Acad. Nat. Sci. Phil. 1:181. 1847.

Twining or prostrate herb; stems slender, to 8 m long, retrorsely hispid; leaves mostly 3-lobed, or the uppermost sometimes unlobed, to 15 cm long, rounded to cordate at the base, pubescent, the lower surface also with resinous glands, the lobes serrate, acute to acuminate at the apex, the petioles to 7 cm long, usually pubescent, the stipules ovate-lanceolate, to 2 cm long, reflexed; staminate flowers in axillary panicles to 10 cm long, with 5 free sepals and 5 free stamens with eglandular anthers; pistillate flowers in an elongated spike, borne two together in the axil of a bract, the bracts eciliate, resinous; achenes ovate, flat, enclosed by the inflated bracts, grouped into an ovoid cluster up to 6 cm long.

COMMON NAME: Common Hops.

HABITAT: Fencerows, thickets, dry, wooded slopes.

RANGE: New Brunswick to Montana, south to California, Arizona, and North Carolina; Europe.

ILLINOIS DISTRIBUTION: Occasional throughout the state.

Although some botanists distinguish the North American plants from the European *H. lupulus*, calling them *H. americanus*, I do not believe there is much justification for doing this.

Hops are used in the brewing industry, where the resinous bracts subtending the pistillate flowers impart a bitter taste and prevent bacterial decay.

Humulus lupulus is found in disturbed areas as well as under undisturbed, natural conditions.

The flowers are borne in July and August.

2. **Humulus japonicus** Sieb. & Zucc. Fl. Jap. Fam. Nat. in Abh. Acad. Muench. 2:89. 1846. *Fig. 43.*

Twining or prostrate herb; stems slender, to 10 m long, retrorsely hispid; leaves mostly 5- to 7-lobed, or the uppermost sometimes

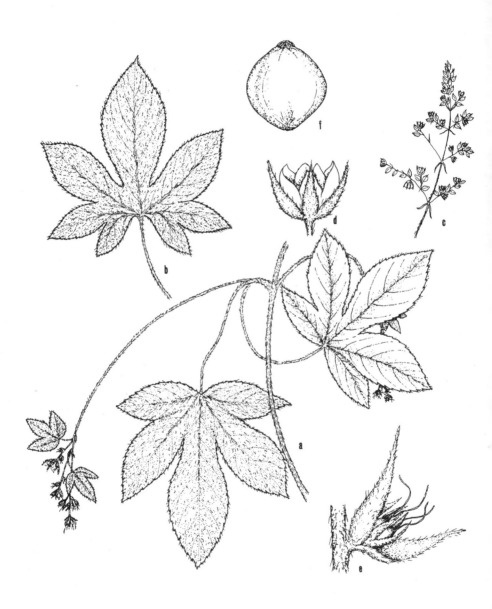

43. *Humulus japonicus* (Japanese Hops). *a*. Leafy branch, with pistillate flowers, ×½. *b*. Leaf variation, ×½. *c*. Staminate panicle, ×½. *d*. Staminate flower, ×5. *e*. Pistillate inflorescence, ×5. *f*. Seed, ×3.

unlobed or 3-lobed, to 15 cm long, rounded to cordate at the base, pubescent, scabrous, the lower surface not with resinous glands, the lobes serrate, acute to acuminate at the apex, the petioles to 15 cm long, usually pubescent, the stipules ovate-lanceolate, to 2 cm long, reflexed; staminate flowers in narrow axillary panicles to 25 cm long, with 5 free sepals and 5 free stamens with glandular anthers; pistillate flowers in a rounded spike, borne two together in the axil of a narrow bract, the bracts hispid and ciliate, not resinous; achenes ovate, flat, not enclosed by the bracts, grouped into a nearly globose cluster up to 4 cm in diameter.

COMMON NAME: Japanese Hops.

HABITAT: Waste ground.

RANGE: Native of Asia; naturalized in the northeastern United States.

ILLINOIS DISTRIBUTION: Occasional and scattered in Illinois.

Japanese hops is sometimes grown as a garden ornamental, but has no value in the brewing industry because it lacks resinous glands on the pistillate bracts. Should the binomial *Humulus scandens* (Lour.) Merr., based on *Antidesma scandens* Lour., prove to be identical with *H. japonicus*, then that binomial would have priority over *H. japonicus*.

The differences between *H. lupulus* and *H. japonicus* are summarized below.

H. lupulus	*H. japonicus*
Leaves mostly 3-lobed.	Leaves mostly 5- to 7-lobed.
Lower leaf surface resinous.	Lower leaf surface not resinous.
Petioles shorter than the blades.	Petioles as long as or longer than the blades.
Anthers eglandular.	Anthers glandular.
Pistillate bracts resinous.	Pistillate bracts not resinous.
Pistillate bracts eciliate.	Pistillate bracts ciliate.
Bracts enclosing achenes.	Bracts not enclosing achenes.

Japanese hops flowers from July to September.

5. *Cannabis* L.–Hemp

Dioecious, annual herb; leaves alternate, lobed, the stipules persistent; flowers unisexual, the staminate in axillary racemes or pan-

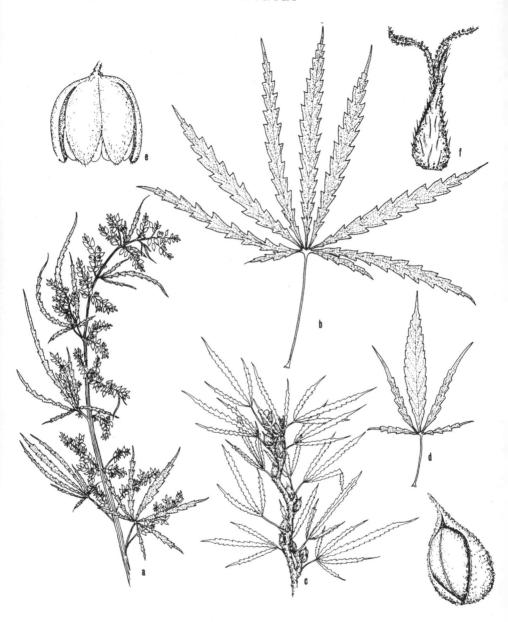

44. *Cannabis sativa* (Marijuana). *a.* Leafy branch, with staminate inflorescences, ×¼. *b.* Leaf from staminate plant, ×1. *c.* Leafy branch, with pistillate
inflorescences, ×¼. *d.* Leaf from pistillate plant, ×½. *e.* Staminate flower,
×5. *f.* Pistil, ×5. *g.* Fruit, ×7½.

icles, the pistillate solitary in the axils; calyx 5-parted or entire; stamens 5, free; ovary superior, 1-locular, with 1 pendulous ovule; styles 2-cleft; fruit an achene.

Only the following species comprises the genus.

1. Cannabis sativa L. Sp. Pl. 1027. 1753. *Fig. 44.*

Coarse, annual herb; stems erect, simple or branched, rough-pubescent, to 2.5 m tall; leaves palmately 3- to 7-lobed, to 15 cm long, rough-pubescent on both surfaces, the lobes acuminate, linear-lanceolate, serrate, the petioles rough-pubescent; staminate flowers in narrow panicles to 10 cm long, on slender pedicels, the 5 sepals nearly free, the 5 stamens pendulous; pistillate flowers in erect, leafy bracted spikes to 2 cm long, the calyx entire, enclosed by the pubescent bract; stigmas 2, protruding above the bracts; achenes compressed, ovoid-oblongoid, to 4 mm long, crustaceous, adnate to the persistent bract.

COMMON NAMES: Marijuana; Hemp.

HABITAT: Floodplains, roadsides.

RANGE: Native of Asia; adventive throughout most of North America.

ILLINOIS DISTRIBUTION: Occasional throughout the state.

This species occurs as an occasional waif along roads and railroads. Occasionally it may grow in large stands. It is illegal to grow this species because of its drug properties.

The dried pistillate inflorescence and upper leaves are the source of marijuana. The fibrous inner bark is important in the making of ropes, twine, and other cordage.

Cannabis sativa flowers from June to October.

URTICACEAE–NETTLE FAMILY

Herbs (in our area), shrubs, or trees; leaves alternate or opposite, simple, with stipules; flowers unisexual or perfect, monoecious or dioecious, variously arranged; calyx 2- to 5-lobed, or the sepals free; petals absent; stamens 2–5, free; ovary superior, 1-locular, with one basal ovule; style 1; fruit an achene.

The Urticaceae contains about 40 genera and 600 species distributed over much of the World.

Probably the most economically important member of the

family is *Boehmeria nivea* Gaud. of China, Japan, and Malaysia, the source of ramie fibers.

KEY TO THE GENERA OF Urticaceae IN ILLINOIS

1. Leaves opposite (a few alternate leaves may occur in *Boehmeria*) __ 2
1. Leaves all alternate _____ 4
   2. Leaves and usually the stems pubescent _____ 3
   2. Leaves and stems glabrous _____ 3. *Pilea*
3. Pistillate flowers with free sepals; stinging hairs often present _____
   _____ 1. *Urtica*
3. Pistillate flowers with united sepals; stinging hairs never present ____
   _____ 2. *Boehmeria*
   4. Leaves serrate; stinging hairs present _____ 4. *Laportea*
   4. Leaves essentially entire; stinging hairs absent ____ 5. *Parietaria*

1. *Urtica* L.–Nettle

Annual or perennial herbs, often with stinging hairs; leaves opposite, simple, toothed, stipulate; flowers unisexual, monoecious or dioecious; staminate flowers in racemes or spikes, the calyx deeply 4-lobed, the stamens 4, free; pistillate flowers in racemes or spikes, the calyx unequally 4-lobed; fruit an achene enclosed by the persistent calyx.

Urtica is a genus of about thirty species found throughout much of the World. The genus differs from other genera of Urticaceae in Illinois by its stinging hairs and opposite leaves.

Hermann (1946) has studied the *U. dioica* complex.

KEY TO THE SPECIES OF Urtica IN ILLINOIS

1. Perennials often attaining a height of 1 m or more; inflorescence 2 cm long or longer _____ 1. *U. dioica*
1. Annuals never attaining a height of 1 m; inflorescence less than 2 cm long _____ 2
   2. Leaves dentate, tapering to the base _____ 2. *U. urens*
   2. Leaves crenate, some of them cordate at the base _____
   _____ 3. *U. chamaedryoides*

1. **Urtica dioica** L. Sp. Pl. 984. 1753. *Fig.* 45.

Urtica gracilis Ait. Hort. Kew. 3:341. 1789.

Urtica procera Muhl. ex Willd. Sp. Pl. 4:353. 1805.

Urtica dioica L. var. *procera* (Muhl.) Wedd. Mon. Fam. Urt. 78. 1856.

45. *Urtica dioica* (Stinging Nettle). *a.* Leafy branch, with pistillate inflorescences below, fruits above, ×1. *b.* Leaf variation, ×1. *c.* Staminate flower, ×10. *d.* Pistillate flower, ×10. *e.* Fruit, ×10.

Perennial monoecious or dioecious herbs; stems erect, simple or branched, cinereous-pubescent to setose, sometimes glabrous or nearly so below, to 2 m tall; leaves lanceolate to ovate, acute to acuminate at the apex, cordate to rounded at the base, coarsely serrate, 3- to 5-nerved from the base, glabrous, pubescent, or setulose on both surfaces, to 15 cm long, up to half as broad, the petioles to 5 cm long, glabrous, pubescent, or setulose, the stipules lanceolate, usually pubescent, to 1.5 cm long; flowers monoecious or dioecious, in cymose-paniculate inflorescences; calyx deeply 4-lobed, pubescent; stamens 4, free; achene ovoid, pubescent, about twice as long as the subpersistent calyx.

COMMON NAME: Stinging Nettle.

HABITAT: Rich woods; moist waste ground.

RANGE: Newfoundland to Alaska, south to Oregon, New Mexico, Louisiana, and North Carolina.

ILLINOIS DISTRIBUTION: Occasional to common in the northern half of the state, rare in the southern half.

I am combining *U. gracilis* and *U. procera* with *U. dioica* because the differences listed by Fernald (1950) to distinguish these three entities seem very questionable.

Most of the Illinois material had cinereous-pubescent stems and leaves, with only a small number of stinging bristles. Both monoecious and dioecious plants have been seen in Illinois.

Urtica dioica flowers from June to September.

2. **Urtica urens** L. Sp. Pl. 984. 1753. *Fig. 46.*

Annual herbs; stems ascending to erect, simple or branched, to 50 cm long, with numerous stinging bristles; leaves elliptic to ovate, acute at the apex, cuneate at the base, coarsely dentate, glabrous or puberulent, 3- to 5-nerved from the base, to 6 cm long, up to half as broad, the petioles to 6 cm long, glabrous or nearly so, the stipules lanceolate, to 5 mm long; flowers monoecious, androgynous, borne in oblong spikes from the leaf axils, the spikes usually shorter than the petioles; calyx deeply 4-lobed, in two very unequal pairs, ciliate; stamens 4, free; achene oblongoid, sparsely pubescent to nearly glabrous, scarcely longer than the two longest calyx lobes.

COMMON NAME: Burning Nettle.
HABITAT: Waste areas.
RANGE: Native of Europe and Asia; naturalized in most of the United States.
ILLINOIS DISTRIBUTION: Known only from Champaign County.
This species is more bristly pubescent than any other species of *Urtica* in Illinois.
This species is similar to *U. chamaedryoides*, another annual, but differs by its dentate rather than crenate leaves and by its cuneate rather than cordate leaf bases.
The flowers open from June to September.

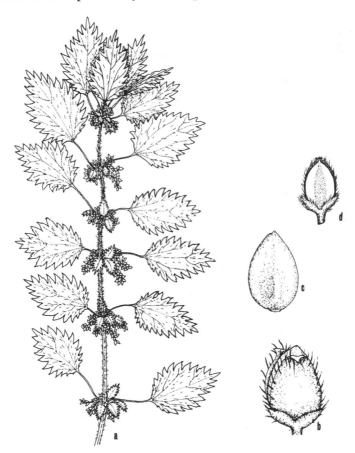

46. Urtica urens (Burning Nettle). *a.* Leafy branch, × ½. *b.* Staminate flower, × 10. *c.* Seed, × 10. *d.* Fruit from front, large sepal removed, × 15.

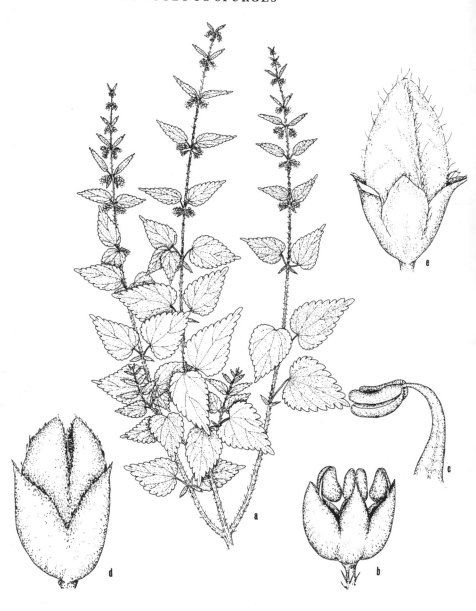

47. *Urtica chamaedryoides* (Nettle). *a.* Leafy branches, × ½. *b.* Staminate flower, × 10. *c.* Stamen, × 17½. *d.* Pistillate flower, × 10. *e.* Fruit, × 10.

3. **Urtica chamaedryoides** Pursh, Fl. Am. Sept. 113. 1814.
Fig. 47.

Annual herbs; stems ascending, simple or branched, sparsely bristly, to 75 cm tall; leaves ovate, obtuse to acute at the apex, mostly cordate at the base, coarsely crenate, mostly pubescent, 3- to 5-nerved from the base, to 5 cm long, to 3.5 cm broad, becoming abruptly smaller toward apex, the petioles to 3 cm long, glabrous or puberulent, the stipules very narrow, to 5 mm long; flowers monoecious, androgynous, in nearly globose spikes from the axils and also terminal; calyx deeply 4-parted, glabrous or puberulent; stamens 4, free; achene oblong-ovoid, sparsely pubescent, twice as long as the subpersistent calyx.

COMMON NAME: Nettle.

HABITAT: Swampy woods, river banks.

RANGE: West Virginia to Oklahoma, south to Texas and Florida; Mexico.

ILLINOIS DISTRIBUTION: Known only from Alexander and Jackson counties.

The first collection of this species from Illinois was made on July 4, 1872, from Grand Tower by G. H. French. Nearly one hundred years later (in 1969), Dan K. Evans collected the same species from apparently the same location.

The flowers bloom from April to June.

2. *Boehmeria Jacq.*–False Nettle

Perennial herbs without stinging hairs; leaves mostly opposite, simple, toothed, stipulate; flowers unisexual, monoecious or dioecious; staminate flowers in racemes or spikes, the calyx deeply 4-lobed, the stamens 4, free; pistillate flowers in racemes or spikes, the calyx tubular, 2- to 4-lobed or entire, enclosing the ovary; fruit an achene enclosed by the persistent calyx.

Boehmeria is a genus of about fifty species native primarily to the tropics. It differs from *Urtica* by the absence of stinging hairs and the tubular pistillate calyx.

Only the following species occurs in Illinois.

1. **Boehmeria cylindrica** (L.) Sw. Prodr. 34. 1788.

Urtica cylindrica L. Sp. Pl. 984. 1753.

Perennial herbs; stems erect, simple or branched, glabrous to rough-pubescent, to 80 cm tall; leaves ovate-lanceolate to ovate, acute to acuminate at the apex, cuneate to rounded at the base, coarsely dentate, smooth to scabrous above, glabrous or pubescent beneath, 3-nerved from the base, to 7 cm long, up to half as broad, the petiole to 9 cm long, glabrous or pubescent, the stipules lance-subulate; flowers monoecious or dioecious, the staminate in interrupted axillary spikes, the pistillate in usually uninterrupted axillary spikes; staminate calyx deeply 4-lobed, pubescent; pistillate calyx tubular, entire or very shallowly 4-toothed; achene ellipsoid, to 2 mm long, enclosed by the persistent calyx, glabrous or beset with hooked hairs.

Two varieties may be distinguished in Illinois.

1. Leaves smooth or slightly scabrous above _____
   _____ 1a. *B. cylindrica* var. *cylindrica*
1. Leaves harshly scabrous above _____
   _____ 1b. *B. cylindrica* var. *drummondiana*

1a. **Boehmeria cylindrica** (L.) Sw. var. **cylindrica** *Fig. 48a–d*.

Urtica capitata L. Sp. Pl. 985. 1753.

Leaves smooth or slightly scabrous above.

COMMON NAME: False Nettle.

HABITAT: Moist woods, marshes.

RANGE: Southern Quebec to southern Ontario, south to Texas and Florida; West Indies.

ILLINOIS DISTRIBUTION: Common throughout the state. The typical variety is a common plant in Illinois, where it usually grows in moist, shaded situations.

In addition to the character given in the key, it also differs from var. *drummondiana* by its usually longer petioles, its flat, unfolded leaves, and its nearly glabrous stems and achenes.

The flowers bloom from July to October.

1b. **Boehmeria cylindrica** (L.) Sw. var. **drummondiana** (Wedd.)
   Wedd. in DC. Prodr. 16 (1):202. 1869. *Fig. 48e*.

Boehmeria drummondiana Wedd. Ann. Sci. Nat. Ser. IV, 1:201. 1854.

Boehmeria cylindrica (L.) Sw. var. *scabra* Porter, Bull. Torrey Club 16:21. 1889.

Boehmeria scabra (Porter) Small, Fl. S. E. U. S. 358. 1903.

Leaves harshly scabrous above.

48. *Boehmeria cylindrica* (False Nettle). *a*. Leafy branch, with pistillate flowers, ×1. *b*. Staminate flower, ×10. *c*. Fruit, ×10. *d*. Seed, ×10. var. *drummondiana*. *e*. Leafy branch, with fruits, ×½.

COMMON NAME: False Nettle.
HABITAT: Bogs and marshes.
RANGE: Massachusetts to Nebraska, south to Texas and Florida.
ILLINOIS DISTRIBUTION: Scattered in Illinois.
This variety grows in more open, sunny situations than var. *cylindrica*. The major differences separating the two varieties are listed under var. *cylindrica*.
The flowers bloom from July to October.

3. Pilea Lindl.–Clearweed

Annual or perennial herbs withut stinging hairs; leaves opposite, simple, toothed, with united stipules; flowers unisexual, monoecious or dioecious, arranged in cymes or glomerules; staminate flowers with the calyx deeply 4-lobed and with 4 free stamens; pistillate flowers with the calyx 3-lobed, with 3 scalelike staminodia and one superior ovary without a style; fruit an achene partly enclosed by the persistent calyx.

There are about two hundred species of *Pilea* primarily found in tropical America.

KEY TO THE SPECIES OF Pilea IN ILLINOIS

1. Achenes green, speckled with purple, averaging 1 mm wide _____
--- 1. *P. pumila*
1. Achenes black, averaging 1.5 mm wide _____ 2. *P. fontana*

1. Pilea pumila (L.) Gray, Man. Bot. ed. 1, 437. 1848. *Fig. 49.*
Urtica pumila L. Sp. Pl. 984. 1753.
Adicea pumila (L.) Raf. ex Torr. Fl. N. Y. 2:223. 1843.
Adicea deamii Lunell, Am. Midl. Nat. 3:10. 1913.
Pilea pumila (L.) Gray var. *deamii* (Lunell) Fern. Rhodora 38:169. 1936.

Annual herb from fibrous roots; stems pellucid, erect or decumbent, simple or branched, glabrous, to 70 cm long; leaves ovate, acute or acuminate at the apex, cuneate or rounded at the base, coarsely crenate, 3-nerved from the base, translucent, sparsely pubescent, to 15 cm long, the petiole to 5 cm long, glabrous; flowers unisexual, the staminate with a 4-lobed calyx, the pistillate with a 3-lobed calyx, the lobes lanceolate; achene ovate, flat, green, with purple speckles, averaging 1 mm wide and about as long, partly enclosed by the whitish or greenish persistent calyx.

49. *Pilea pumila* (Clearweed). *a.* Upper part of plant, with leaves and flowers, × ½. *b.* Fruit, with persistent calyx, × 12½. *c.* Fruit, × 20.

COMMON NAME: Clearweed.

HABITAT: Moist, usually shaded, ground.

RANGE: Quebec to Ontario and Minnesota, south to Texas and Florida.

ILLINOIS DISTRIBUTION: Common throughout the state; probably in every county.

Pilea pumila often grows in extensive numbers in low, moist, shaded soil. It is frequent also on rotted logs which have fallen in swamps.

The green seeds speckled with purple distinguish this species from *P. fontana*. Although var. *deamii* (Lunell) Fern. is sometimes recognized as distinct because of its rounded leaf bases and more numerous teeth on the leaves, it does not seem to me to be worthy of recognition.

The clear stem, to which the plant owes its common name, is an interesting phenomenon.

The flowers are produced from July to September.

2. Pilea fontana (Lunell) Rydb. Brittonia 1:87. 1931. *Fig. 50.*

Adicea fontana Lunell, Am. Midl. Nat. 3:7. 1913.

Adicea opaca Lunell, Am. Midl. Nat. 3:8. 1913.

Pilea opaca (Lunell) Rydb. Brittonia 1:87. 1931.

Annual herb from fibrous roots; stems pellucid, erect or decumbent, simple or branched, glabrous, to 50 cm long; leaves ovate, acute or acuminate at the apex, rounded or nearly so at the base, coarsely crenate, 3-nerved from the base, not translucent, glabrous or sparsely pubescent, to 10 cm long, the petiole to 4 cm long, glabrous; flowers unisexual, the staminate with a 4-lobed calyx, the pistillate with a 3-lobed calyx; achenes ovate, flat, black, averaging 1.5 mm long and broad, partly enclosed by the usually purplish, persistent calyx.

COMMON NAME: Black-seeded Clearweed.

HABITAT: Moist soil.

RANGE: New York to North Dakota, south to Nebraska, Illinois, and Florida.

ILLINOIS DISTRIBUTION: Occasional and scattered in Illinois.

Pilea opaca, another black-seeded plant, has been reported from Illinois, but the characters of *P. opaca* are not distinct enough from those of *P. fontana* to merit recognition of *P. opaca*. The leaves of *P. fontana* are not translucent as are those of *P. pumila*.

This species flowers from July to September.

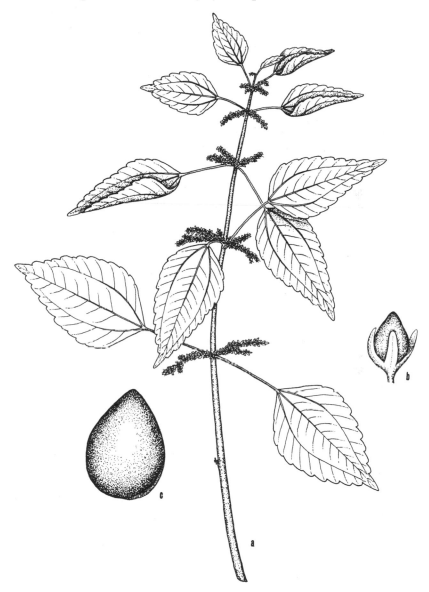

50. *Pilea fontana* (Black-seeded Clearweed). *a.* Upper part of plant, with leaves and flowers, × ½. *b.* Fruit, with persistent calyx, × 10. *c.* Fruit, × 20.

51. *Laportea canadensis* (Wood Nettle). *a*. Leafy branch, with inflorescences, ×1. *b*. Staminate flower, ×15. *c*. Pistillate flowers, ×10. *d*. Fruit, ×15.

4. Laportea Gaud.–Wood Nettle

Perennial herbs with stinging hairs; leaves alternate, simple, toothed, stipulate; flowers unisexual, monoecious or dioecious, arranged in compound cymes; staminate flowers with 5 sepals and 5 stamens; pistillate flowers with 4 sepals in 2 unequal pairs and a superior ovary with a single style; achene reflexed.

Laportea is a genus of about two dozen species, primarily native to the tropics. This is the only Illinois genus of Urticaceae with stinging hairs and alternate leaves.

Only the following species occurs in Illinois.

1. **Laportea canadensis** (L.) Gaud. Bot. Voy. Freyc. 498. 1826.
   Fig. 51.

Urtica canadensis L. Sp. Pl. 985. 1753.
Urtica divaricata L. Sp. Pl. 985. 1753.
Urticastrum divaricatum (L.) Kuntze, Rev. Gen. Pl. 635. 1891.

Perennial herbs; stems erect, simple or branched, beset with stinging hairs, to 2 m tall; leaves ovate, acute to acuminate at the apex, cuneate at the base, sharply serrate, 3-nerved from the base, glabrous or bristly-hairy, to 15 cm long, to 10 cm broad, the petioles to 10 cm long, setose; flowers unisexual, borne in large compound cymes, the staminate with 5 nearly free sepals and 5 stamens, the pistillate with 4 unequal sepals and a slender superior ovary; achene ovate, compressed, glabrous, to 3 mm long, reflexed at maturity.

COMMON NAME: Wood Nettle.

HABITAT: Moist soil.

RANGE: Nova Scotia to Manitoba, south to Oklahoma and Florida.

ILLINOIS DISTRIBUTION: Common throughout the state. This species bears numerous, coarse, stinging hairs which break off when touched. The substance contained in the hairs causes a very uncomfortable sensation when contacting the skin. Since the wood nettle grows in dense thickets in low woods, it causes a formidable barrier to anyone wishing to cross through an area where this species grows.

The wood nettle flowers from June to September.

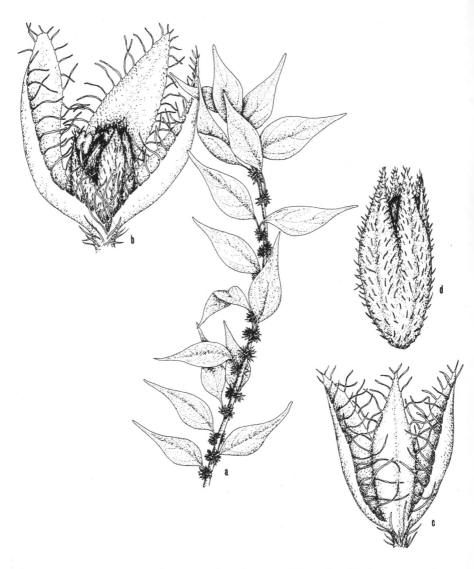

52. *Parietaria pensylvanica* (Pellitory). *a*. Leafy branch, with flowers, ×1. *b*. Perfect flower, subtended by bracts, ×20. *c*. Cluster of bracts, ×20. *d*. Achene enclosed by calyx, ×40.

5. Parietaria L.–Pellitory

Annual or perennial herbs without stinging hairs; leaves alternate, simple, entire, with no stipules; flowers staminate, pistillate, or perfect, monoecious, borne in axillary clusters; staminate flowers with a deeply 4-parted calyx and 4 stamens; pistillate and perfect flowers with a tubular, 4-lobed calyx and a superior ovary; achene enclosed by the subpersistent calyx.

There are about eight species of *Parietaria* found in much of the warm regions of the World.

Only the following species occurs in Illinois.

1. Parietaria pensylvanica Muhl. ex Willd. Sp. Pl. 4(2):955. 1806. *Fig. 52.*

Weak annual from a small tuft of fibrous roots; stems erect or reclining, simple or branched, puberulent, to 50 cm tall; leaves membranous, oblong-lanceolate to lance-ovate, acuminate at the apex, cuneate at the base, entire, 3-nerved from the base, to 7 cm long, to 2 cm broad, with rough punctations on the upper surface, the pubescent petioles to 2.5 cm long; flowers in sessile glomerules from the leaf axils, subtended by linear bracts at least twice as long as the flowers; achene ovoid, glabrous or nearly so, to 1 mm long.

COMMON NAME: Pellitory.

HABITAT: In shade beneath cliffs and along buildings.

RANGE: Quebec to British Columbia, south to Nevada, Texas, and Florida; Mexico.

ILLINOIS DISTRIBUTION: Common throughout the state. This weak, obscure species is often overlooked in Illinois because of its low stature and greenish flowers. It is found in both "natural-looking" habitats such as beneath sandstone cliff overhangs to disturbed areas such as around buildings.

The pellitory flowers from May to September.

Order Rhamnales

In the classification system followed here, the Rhamnales is composed of two families, both of which occur in Illinois. Cronquist's (1968) treatment of the Rhamnaceae and Elaeagnaceae is very different from that of Thorne. Cronquist places the Rhamnaceae, Vitaceae, and the tropical Leeaceae in the Rhamnales, and groups the Elaeagnaceae and Proteaceae in the order Proteales.

RHAMNACEAE–BUCKTHORN FAMILY

Trees or shrubs, sometimes climbing, sometimes with thorns; leaves alternate, simple, stipulate; flowers usually perfect, rarely unisexual and dioecious, variously arranged; sepals 4- to 5-lobed; petals (4–) 5, free, sometimes absent; stamens (4–) 5, free, borne from a disk, the disk fleshy; pistil 1, the ovary superior, 2- to 5-locular, with 1 (–2) ovules per locule; fruit a berry, drupe, capsule, or samara.

This family is composed of about forty-five genera and nearly six hundred species found throughout most of the World.

In addition to the genera enumerated below, the genus *Zizyphus* (the jujube tree) is sometimes planted in Illinois as an ornamental.

KEY TO THE GENERA OF Rhamnaceae IN ILLINOIS

1. Woody vines _____ 1. *Berchemia*
1. Erect trees or shrubs _____ 2
    2. Leaves 3-veined from the base; fruit a capsule ____ 2. *Ceanothus*
    2. Leaves pinnately veined; fruit a drupe _____ 3. *Rhamnus*

1. Berchemia Neck.–Supple-jack

Climbing woody shrub without thorns; leaves alternate, simple, pinnately nerved; flowers perfect, in panicles; calyx 5-lobed; petals 5, free; stamens 5, free; disk filling the calyx tube but not attached to the ovary; pistil 1, the ovary superior but covered by the disk; fruit a 1-seeded drupe.

Ten species make up the genus, with all but ours native to Asia and Africa.

Only the following species occurs in Illinois.

1. **Berchemia scandens** (Hill) K. Koch, Dendr. 1:602. 1869. *Fig. 53.*

Rhamnus scandens Hill, Hort. Kew. 453. 1768.

High-climbing woody vine; stems terete, slender, tough, glabrous; buds lanceoloid, glabrous, closely appressed to the stem; leaf scars round, elevated, with one bundle scar; leaves ovate-oblong, obtuse to acute at the apex, rounded at the base, serrulate or undulate, glabrous on both surfaces, dark green above, paler below, to 5 cm long, to 2.5 cm broad, the petioles to 1 cm long; flowers to 3 mm broad, greenish-white, perfect, in axillary or terminal panicles; calyx 5-lobed, the lobes acute; petals 5, acute; stamens 5, shorter than the petals; drupe ellipsoid, blue, to 8 mm long, 1-seeded.

COMMON NAME: Supple-jack.

HABITAT: In pine plantation (in Illinois).

RANGE: Virginia to southern Missouri, south to Texas and Florida.

ILLINOIS DISTRIBUTION: Known only from Pope County, where it was first collected in 1969.

Although this species has been attributed to Illinois by Gleason (1952) and other writers, the report is based on a specimen in the herbarium of the New York Botanical Garden collected at Pawpaw Junction, which is in Missouri rather than Illinois.

However, *Berchemia scandens* has been collected at the edge of a pine plantation in Pope County in 1969, by John Schwegman.

Supple-jack flowers from April to June.

2. Ceanothus L.

Shrubs, sometimes with thorns; leaves alternate, simple, 3-nerved from the base, flowers perfect, in terminal or axillary panicles or corymbs, on pedicels colored like the petals; calyx 5-lobed; petals 5, free, cucullate, attached beneath the disk; stamens 5, free; ovary sunken in the disk, the style 3-cleft; fruit 3-lobed, dry, dehiscing into three nutlets at maturity.

Ceanothus is a genus of about fifty species native to North America. Many species occur in the Pacific regions of the United States. Several species are grown as ornamentals.

53. *Berchemia scandens* (Supple-jack). *a.* Leafy branch, with inflorescences, × 1. *b.* Flower, with front calyx lobe pulled down and front two petals removed, × 12½. *c.* Fruit, × 7½.

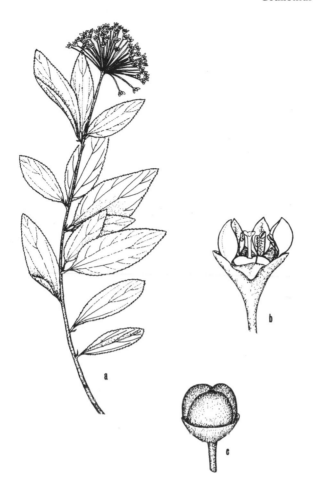

54. *Ceanothus ovatus* (Inland New Jersey Tea). *a.* Leafy branch, with inflorescence, × ¾. *b.* Flower, × 7½. *c.* Fruit, × 7½.

KEY TO THE SPECIES OF Ceanothus IN ILLINOIS

1. Leaves elliptic to elliptic-lanceolate, obtuse to subacute at the apex _____ 1. *C. ovatus*
1. Leaves ovate to ovate-oblong, acute to acuminate (obtuse in var. *pitcheri*) _____ 2. *C. americanus*

1. Ceanothus ovatus Desf. Hist. Arb. 2:381. 1809. *Fig. 54.*

Ceanothus ovalis Bigel. Fl. Bost., ed. 2, 92. 1824.

Shrub to 1 m tall, without thorns; stems erect, branched, puberulent; buds ovoid, acute, pubescent; leaf scars half-round, elevated, with usually one bundle trace; leaves elliptic to elliptic-lanceolate, obtuse to subacute at the apex, cuneate to rounded at the base, serrate, 3-nerved from the base, glabrous or sparsely pilose above and below, to 6 cm long, to 2.5 cm broad, on usually glabrous petioles to 5 mm long; flowers numerous, white, mostly in terminal corymbs, the peduncles short, the pedicels to 1.5 cm long; calyx 5-lobed, the lobes broadly ovate, incurved, white; petals 5, free, white, clawed, the apex cucullate; capsules three-lobed, black, to 5 mm high.

COMMON NAME: Inland New Jersey Tea.

HABITAT: Low dunes, sandy soil.

RANGE: Quebec to Manitoba, south to Texas and Georgia.

ILLINOIS DISTRIBUTION: Known only from Cook, JoDaviess, Lake, Whiteside, and Winnebago counties. Despite the broad range of this species, it is a very rare plant in Illinois, where it is confined to the northern tip of the state.

Swink (1974) reports the inland New Jersey tea as frequent in only one locality in northeastern Illinois, at the south end of Illinois Beach State Park in Lake County. At this station, Swink lists some of the associated species as *Arenaria stricta, Comandra richardsiana, Hypericum kalmianum, Liatris aspera,* and *Potentilla fruticosa.*

Shinners (1951) and Brizicky (1964) have indicated that this species should be called *C. herbaceus.*

Ceanothus ovatus is best distinguished from *C. americanus* by its narrower leaves.

Brendel (1859) is the first to report this species from Illinois, calling it *C. ovalis.*

This species flowers during May and June.

2. Ceanothus americanus L. Sp. Pl. 195. 1753.

Shrub to 1.5 m tall, without thorns; stems erect, branched, glabrous or pubescent; buds ovoid, acute, pubescent; leaf scars half-round, elevated, with usually one bundle trace; leaves ovate to ovate-oblong, obtuse to acute to acuminate at the apex, rounded at the base, serrate, 3-nerved from the base, glabrous or sparsely pilose or velvety, to 10 cm long, to 6 cm broad, usually considerably smaller, on usually glabrous petioles to 1 cm long; flowers numerous, white, in terminal and axillary thyrses, the peduncles slender, the white pedicels to 1.5 cm long; calyx 5-lobed, the lobes deltoid, incurved, white; petals 5, free, white, clawed, the apex cucullate; capsules 3-lobed, black, to 5 mm high.

Two varieties may be distinguished in Illinois.

1. Leaves acute to acuminate at the apex, glabrous or nearly so on the upper surface, sparsely pubescent on the lower surface _____
_____ 2a. *C. americanus* var. *americanus*
1. Leaves obtuse at the apex, pilose on the upper surface, velvety on the lower surface _____ 2b. *C. americanus* var. *pitcheri*

2a. Ceanothus americanus L. var. **americanus** *Fig. 55a–d*.

Ceanothus intermedius Pursh, Fl. Am. Sept. 1:167. 1814.
Ceanothus americanus L. var. *intermedius* (Pursh) K. Koch, Hort. Dendr. 206. 1853.

Leaves acute to acuminate at the apex, glabrous or nearly so on the upper surface, sparsely pubescent on the lower surface.

COMMON NAME: New Jersey Tea.
HABITAT: Woods, thickets.
RANGE: Quebec to Manitoba, south to Louisiana and Florida.
ILLINOIS DISTRIBUTION: Occasional throughout the state.

This is the common variety of *Ceanothus americanus* in Illinois. When in flower, the shrub is extremely attractive.

Fernald (1950) and others subdivide this variety into typical var. *americanus* and var. *intermedius*. Variety *intermedius* is said to differ by its smaller leaves, inflorescences, and fruiting clusters, but intergradation in all three characters renders separation impossible.

The leaves have been a source of tea in the past.
The flowers bloom from June to August.

2b. Ceanothus americanus L. var. **pitcheri** Torr. & Gray, Fl. N. Am. 1:264. 1838. *Fig. 55e*.

Leaves obtuse at the apex, pilose on the upper surface, velvety on the lower surface.

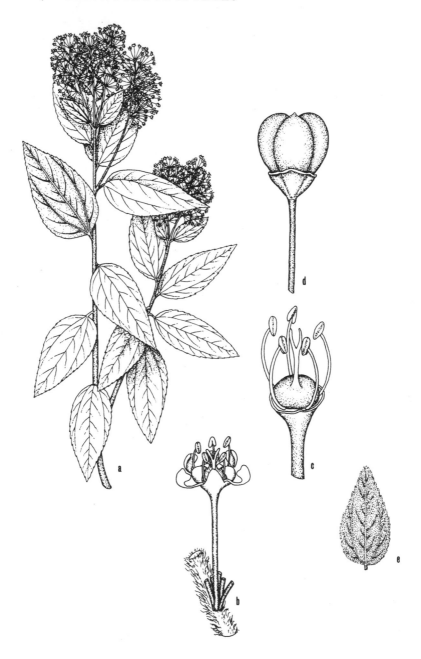

55. *Ceanothus americanus* (New Jersey Tea). *a.* Upper part of plant with flowers, × ½. *b.* Flower, × 6½. *c.* Flower, with perianth removed, × 15. *d.* Capsule, × 3. var. *pitcheri. e.* Leaf, × ½.

COMMON NAME: New Jersey Tea.
HABITAT: Woods.
RANGE: Indiana to Kansas, south to Georgia and Texas.
ILLINOIS DISTRIBUTION: Not common, but scattered throughout the state.
Variety *pitcheri* has obtuse leaf apices similar to those found in *C. ovatus*. The velvety lower leaf surface of var. *pitcheri* easily distinguishes these two, however. The flowers bloom from June to August.

3. Rhamnus L.–Buckthorn

Trees or shrubs, sometimes with thorns; leaves alternate, simple, pinnately nerved; flowers perfect or unisexual, monoecious or dioecious, arranged in axillary cymes, panicles, or racemes; calyx campanulate, 4- to 5-lobed; petals 4–5, free, clawed, sometimes cucullate, or petals absent; stamens 4–5; disk fleshy, not attached to the ovary; style 3- to 4-cleft; drupe berrylike, with 2–4 nutlets.

Rhamnus is composed of about one hundred species, mostly native to the temperate Northern Hemisphere.

Several species are grown as ornamentals. The fruits of some species have cathartic properties.

KEY TO THE SPECIES OF Rhamnus IN ILLINOIS

1. Low shrub less than 1 m tall; petals absent _____ 1. *R. alnifolia*
1. Shrubs or trees, at maturity over 1 m tall; petals present _____ 2
  2. Some of the branchlets ending in a spine; nutlets with a narrow groove _____ 3
  2. None of the branchlets ending in a spine; nutlets without a groove, or the groove broad and open _____ 4
3. Leaves ovate to elliptic, dull on the upper surface _ 2. *R. cathartica*
3. Leaves obovate to oblong, shiny on the upper surface _____
  _____ 3. *R. davurica*
  4. Winter buds with scales; flowers appearing with the leaves; nutlets with a broad, open groove _____ 4. *R. lanceolata*
  4. Winter buds without scales; flowers appearing after the leaves; nutlets without a groove _____ 5
5. Umbels peduncled; leaves more or less acute; most or all the petioles 5 mm long or longer; pedicels pubescent _____ 5. *R. caroliniana*
5. Umbels sessile; leaves more or less obtuse; most or all the petioles less than 5 mm long; pedicels glabrous or nearly so _____ 6. *R. frangula*

1. **Rhamnus alnifolia** L'Hér. Sert. Angl. 3. 1788. *Fig. 56*.

Shrub to 75 cm tall, without thorns; twigs slender, red or brown, puberulent; buds ovoid, subacute, to 5 mm long; leaf scars half-round, slightly elevated, with 3 bundle scars; leaves elliptic to oval, obtuse to acute at the apex, rounded to cuneate at the base, serrulate, glabrous or nearly so, to 10 cm long, up to half as broad, on glabrous petioles to 1.5 cm long; flowers mostly unisexual, dioecious, 1–3 per leaf axil; calyx campanulate, with 5 acute, deltoid, glabrous lobes; petals absent; stamens 5, free; drupes globose, black, to 6 mm in diameter, with 3 shallowly grooved, flat nutlets.

COMMON NAME: Alder-leaved Buckthorn.

HABITAT: Bogs, wooded swamps.

RANGE: Newfoundland to British Columbia, south to California, Nebraska, northern Illinois, and West Virginia.

ILLINOIS DISTRIBUTION: Confined to the northern one-half of the state and Richland County, but much less common than the map indicates due to destruction of suitable habitats. Swink (1974) reports it to be more common in springy calcareous situations than in bogs.

This is the lowest growing species of *Rhamnus* in Illinois, never reaching a height of one meter. It is also the only *Rhamnus* in Illinois which lacks petals.

The alder-leaved buckthorn flowers from May to July.

2. **Rhamnus cathartica** L. Sp. Pl. 193. 1753. *Fig. 57*.

Shrub or small tree to 6 m tall, with spine-tipped branches; twigs slender, gray or brown, glabrous; buds lance-ovoid, acute, to 5 mm long; leaf scars half-elliptic, slightly elevated, with 3 bundle scars; leaves elliptic to ovate, obtuse to acute at the apex, rounded or subcordate at the base, dull on the upper surface, crenulate, glabrous, to 5.5 cm long, to 2.5 cm broad, on glabrous petioles to 1 cm long; flowers mostly unisexual, dioecious, greenish, to 2 mm across, 1–4 per leaf axil; calyx campanulate, with 4 acute lobes; petals 4, free, linear-lanceolate; stamens 4, free; drupe globose, black, to 8 mm in diameter, with 3–4 deeply grooved nutlets.

56. *Rhamnus alnifolia* (Alder-leaved Buckthorn). *a.* Branch, with leaves and fruits, ×½. *b,c.* Flowers, ×5. *d.* Fruit, ×4. *e.* Seed, ×7½.

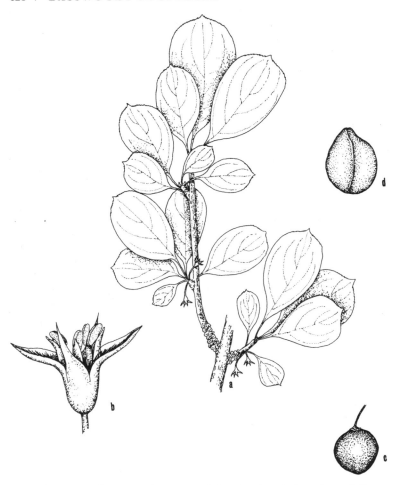

57. *Rhamnus cathartica* (Common Buckthorn). *a*. Leafy branch, with flowers, ×¾. *b*. Staminate flower, ×12½. *c*. Fruit, ×2½. *d*. Nutlet, ×5.

COMMON NAME: Common Buckthorn.

HABITAT: Disturbed woodlands and waste areas.

RANGE: Native of Europe; naturalized from Quebec to Minnesota, south to Missouri and Virginia.

ILLINOIS DISTRIBUTION: Occasional in the northern three-fifths of the state.

Most parts of this plant contain a strongly purgative substance.

Some of the leaves, because of crowded internodes, appear to be opposite.

This species is similar to *R. davurica* except that it has dull, ovate to elliptic leaves.

In northeastern Illinois, this is an obnoxious weed in disturbed woods and along fences. Its spread is partly attributed to seed dispersal by birds.

The common buckthorn flowers during May and June.

3. **Rhamnus davurica** Pall. Fl. Ross. 2, t. 61. 1788. *Fig. 58.*

Small tree to 10 m tall, with spine-tipped branches; twigs slender, gray or brown, glabrous; buds lanceoloid, acute, to 4 mm long; leaf scars half-elliptic, slightly elevated, with 3 bundle scars; leaves oblong to obovate, obtuse to acute at the apex, rounded or subcordate at the base, shiny on the upper surface, serrate, glabrous, to 7 cm long, to 3.5 cm broad, on glabrous petioles to 1 cm long; flowers mostly unisexual, dioecious, greenish-yellow, to 2.5 mm across, few in the axils of the leaves; calyx campanulate, with 4 acute lobes; petals 4, free, linear-lanceolate; stamens 4, free; drupe globose, black, to 8 mm in diameter, with 3–4 deeply grooved nutlets.

COMMON NAME: Buckthorn.

HABITAT: Cultivated soil.

RANGE: Native of Asia; rarely adventive in North America.

ILLINOIS DISTRIBUTION: Known only from DuPage County.

This species is very similar to *R. cathartica*, but differs by its shiny, obovate to oblong leaves.

The specific epithet is sometimes spelled *dahurica.*

This species flowers during May and June.

4. **Rhamnus lanceolata** Pursh, Fl. Am. Sept. 166. 1814. *Fig. 59.*

Rhamnus lanceolata Pursh var. *glabrata* Gl. Phytologia 2:288. 1947.

Shrub to 5 m tall, without thorns; twigs slender, gray, puberulent; buds ovoid, acute, to 5 mm long; leaf scars half-round, slightly elevated, with 3 bundle scars; leaves lanceolate to lance-ovate, obtuse to acute to acuminate at the apex, rounded to cuneate at the base, serrulate, glabrous or nearly so on the upper surface, glabrous or puberulent on the lower surface, to 8 cm long, to 2.5 cm broad, on

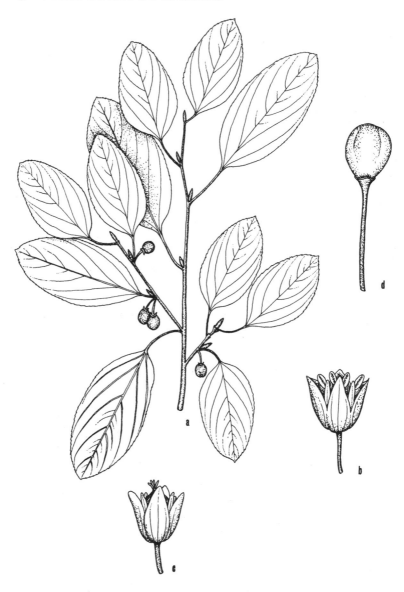

58. *Rhamnus davurica* (Buckthorn). *a.* Leafy branch with fruits, ×½. *b.* Staminate flower, ×5. *c.* Pistillate flower, ×5. *d.* Fruit, ×3.

puberulent to nearly glabrous petioles to 1 cm long; flowers bisexual, of two kinds on separate plants, one with an included style, one with an exserted style, in axillary clusters of 1–3 per leaf axil, green-

59. *Rhamnus lanceolata* (Lance-leaved Buckthorn). *a.* Leafy branch, with flowers, ×1. *b.* Staminate flower, ×10. *c.* Pistillate flower, ×10. *d.* Fruit, ×3. *e.* Nutlet, ×6.

ish-yellow, to 3 mm across; calyx campanulate, with 4 acute lobes; petals 4, free, bifid; stamens 4, free; drupe globose, black, to 8 mm in diameter, with 2 deeply grooved nutlets.

COMMON NAME: Lance-leaved Buckthorn.

HABITAT: Riverbanks, bluffs, calcareous fens.

RANGE: Pennsylvania to southern Wisconsin to Nebraska, south to Texas and Alabama.

ILLINOIS DISTRIBUTION: Rare to occasional in the northern two-thirds of the state; rare in southern Illinois.

The flowers of this species appear as the leaves begin to unfold. This character, along with the grooved nutlets and the scaly winter buds, distinguishes *R. lanceolata* from *R. caroliniana* and *R. frangula*.

Gleason's var. *glabrata*, with the leaves glabrous or nearly so on the lower surface, seems scarcely worthy of recognition.

The flowers bloom from April to June.

5. Rhamnus caroliniana Walt. Fl. Car. 101. 1788. *Fig. 60.*

Frangula caroliniana (Walt.) Gray, Gen. 2:178. 1849.

Rhamnus caroliniana Walt. var. *mollis* Fern. Rhodora 12:79. 1910.

Shrub or small tree to 10 m tall, without thorns; twigs slender, red-brown to gray, puberulent to glabrous; buds ovoid, acute, tomentose, without scales, 4–5 mm long; leaf scars oval, slightly elevated, with 3 bundle scars; leaves elliptic to oblong, acute at the apex, rounded to cuneate at the base, obscurely serrulate or crenulate, glabrous or velvety-pubescent on both surfaces, to 15 cm long, to 4 cm broad, on glabrous or pubescent petioles to 2 cm long; flowers perfect, solitary or few-flowered in axillary, pedunculate umbels, greenish, to 2 mm broad; calyx campanulate, 5-lobed, the lobes acute to acuminate, glabrous or puberulent; petals 5, free, bifid; stamens 5, free; drupe globose, red, to 8 mm in diameter, with 2–4 ungrooved nutlets.

COMMON NAME: Carolina Buckthorn.

HABITAT: Woods.

RANGE: Virginia to Nebraska, south to Texas and Florida.

ILLINOIS DISTRIBUTION: Not common; confined to the southern one-fourth of the state.

The lustrous green leaves and the bright red berries make this a most attractive shrub or small tree.

Considerable variation is exhibited in the degree of pubescence of the leaves. Those plants with the lower leaf surface velvety-pubescent have been designated var.

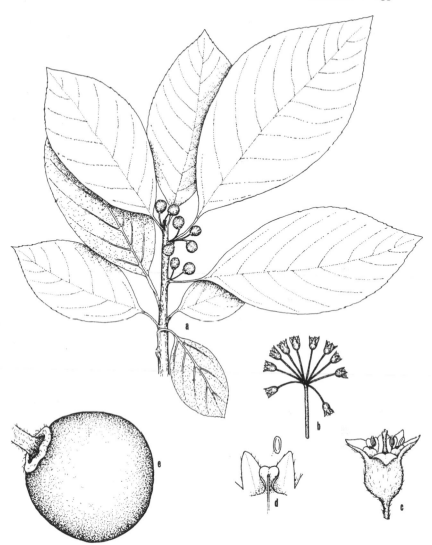

60. *Rhamnus caroliniana* (Carolina Buckthorn). *a.* Leafy branch, with fruits, ×¾. *b.* Inflorescence, ×½. *c.* Staminate flower, ×10. *d.* Anther removed, showing cordate corolla lobe, ×10. *e.* Fruit, ×7½.

mollis. The type for this scarcely justifiable variety was collected at Grand Tower, Jackson County.

The flowers, which appear after the leaves have unfolded, bloom during May and June.

6. Rhamnus frangula L. Sp. Pl. 193. 1753.

Shrub or small tree to 5 m tall, without thorns; twigs slender, gray to brown, puberulent; buds ovoid, acute, tomentose, without scales, 4–5 mm long; leaf scars oval, slightly elevated, with 3 bundle scars; leaves obovate to narrowly lanceolate, obtuse to acute at the apex, rounded to cuneate at the base, entire or sparsely crenulate, glabrous, to 7 cm long, to 4 cm broad, the glabrous petioles to 1 cm long; flowers perfect, solitary or few-flowered in axillary, sessile umbels, greenish, to 2 mm broad; calyx campanulate, 5-lobed, the lobes acute, glabrous; petals 5, free, bifid; stamens 5, free; drupe globose, red at first, becoming black, to 8 mm in diameter, with 3 ungrooved nutlets.

Two varieties occur in Illinois.

1. Leaves obovate _____ 6a. *R. frangula* var. *frangula*
1. Leaves narrowly lanceolate _____ 6b. *R. frangula* var. *angustifolia*

6a. Rhamnus frangula L. var. frangula *Fig. 61a–d*.

Leaves obovate.

COMMON NAME: Glossy Buckthorn.
HABITAT: Disturbed woods and bogs.
RANGE: Native of Europe; naturalized throughout the eastern half of the United States.
ILLINOIS DISTRIBUTION: Occasional in the northern half of the state; also Crawford County.
Rhamnus frangula and *R. caroliniana* belong to section Frangula, differing from other buckthorns by their scaleless winter buds and their perfect flowers.
The bird-disseminated fruits apparently account for the spread of this species into disturbed woods and bogs.
This variety flowers from May to July.

61. *Rhamnus frangula* (Glossy Buckthorn). *a*. Leafy branch, with flowers, ×¾. *b*. Flower, ×12½. *c*. Fruit, ×5. *d*. Nutlet, ×10. var. *angustifolia* (Narrow-leaved Glossy Buckthorn). *e*. Leafy branch, with flowers, ×¼.

6b. **Rhamnus frangula** L. var. **angustifolia** Loud. Arb. & Fruct.
2:537. 1838. *Fig. 61e.*
Leaves narrowly lanceolate.

COMMON NAME: Narrow-leaved Glossy Buckthorn.
HABITAT: Disturbed areas.
RANGE: Native of Europe; naturalized in Illinois.
ILLINOIS DISTRIBUTION: Known only from Cook County.
This variety, in the vegetative conditions, looks more like a species of willow than a buckthorn.
The flowers are borne in June and July.

ELAEAGNACEAE–OLEASTER FAMILY

Trees or shrubs with peltate or stellate scales; leaves alternate or opposite, simple, entire, without stipules; flowers perfect or unisexual, monoecious or dioecious, usually in axillary clusters; calyx usually tubular, 2- to 4-lobed, becoming fleshy in fruit, or sepals 4 and free or nearly so in staminate flowers; petals absent; stamens 4 or 8; disk lobed or annular; ovary 1-locular, 1-ovulate; fruit an achene or nut, surrounded by the fleshy calyx, resembling a drupe.

This family is composed of three genera and about fifty species, native mostly to the northern hemisphere.

Members of the family are usually covered by silvery or golden-brown scales.

KEY TO THE GENERA OF Elaegnanceae IN ILLINOIS

1. Leaves opposite; stamens 8; flowers unisexual _____ 1. *Shepherdia*
1. Leaves alternate; stamens 4; flowers bisexual _____ 2. *Elaeagnus*

1. *Shepherdia Nutt.*–Buffalo-berry

Dioecious shrubs with brown or silvery stellate scales; leaves opposite, entire; flowers unisexual, in axillary clusters or at the nodes of the preceding years; pistillate flowers with a 4-lobed urceolate calyx, an 8-lobed disk, and a 1-locular ovary; staminate flowers with 4 nearly free sepals and 8 stamens; fruit a nut or achene, enclosed by the fleshy, persistent calyx.

Shepherdia is a genus of three North American species. In addition to the species in Illinois, two more occur in the western United States.

Only the following species occurs in Illinois.

1. **Shepherdia canadensis** (L.) Nutt. Gen. 2:240. 1818. *Fig. 62.*

Elaeagnus canadensis L. Sp. Pl. 1024. 1753.

Shrubs to 2 m tall; twigs slender, brown, scurfy, with peltate scales; buds scurfy, obtuse, to 3 mm long; leaf scars half-round, slightly elevated, with 1 bundle scar; leaves opposite, elliptic to oval, obtuse at the apex, rounded or cuneate at the base, entire, sparsely stellate-scurfy and green above, densely silvery-scurfy beneath, intermixed with rusty scales, to 3 cm long, to 2.5 cm broad, on scurfy petioles to 6 mm long; flowers unisexual, solitary or few in the axils or from the nodes of last year's branches, to 5 mm across; staminate flowers with 4 free, lanceolate, acute sepals and 8 free stamens; pistillate flowers with a 4-lobed, urceolate calyx and a large ovary; fruit oval, yellowish-red, to 6 mm long, containing 1 smooth nut.

COMMON NAME: Canadian Buffalo-berry.

HABITAT: Sandy shores of Lake Michigan.

RANGE: Newfoundland to Alaska, south to New Mexico, South Dakota, northern Illinois, and New York.

ILLINOIS DISTRIBUTION: Restricted to Cook and Lake counties.

The Canadian buffalo-berry is distinguished by its silvery-scurfy opposite leaves and its unisexual flowers. This is one of the rarest shrubs in Illinois. Only a few plants occur in sand along Lake Michigan. Vasey first reported it from Illinois in 1861.

The flowers are borne from May to July.

2. **Elaeagnus** L.–Oleaster

Shrubs or small trees with silvery stellate scales or hairs; leaves alternate, entire; flowers perfect, 1–4 per leaf axil; calyx tubular, 4-lobed; petals absent; stamens 4, free, attached to the throat of the calyx tube; ovary 1-locular, with 1 ovule; fruit a nut enclosed by the fleshy persistent calyx, drupelike.

There are about twenty species of *Elaeagnus* native to Asia, Australia, western North America, and Europe. Several species are grown as ornamentals.

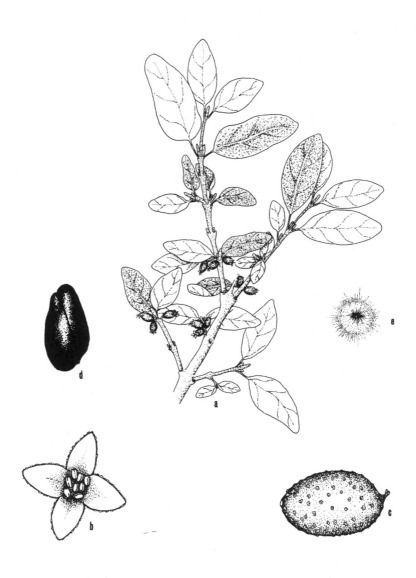

62. *Shepherdia canadensis* (Canadian Buffalo-berry). *a.* Fruiting branch, ×¾. *b.* Staminate flower, ×7½. *c.* Fruit, ×5. *d.* Nutlet, ×6¼. *e.* Hair, ×50.

KEY TO THE SPECIES OF Elaeagnus IN ILLINOIS

1. Fruit yellow; branchlets and leaves with silvery scales only _____
   _____ 1. *E. angustifolia*
1. Fruit pink or red; branchlets and leaves with brown and silvery
   scales _____ 2
   2. Fruits on stalks up to 6 mm long _____ 2. *E. umbellata*
   2. Fruits on stalks 15–22 mm long _____ 3. *E. multiflora*

1. **Elaeagnus angustifolia** L. Sp. Pl. 121. 1753. *Fig. 63.*

Shrub or small tree to 6 m tall; twigs slender, silvery-white; buds scurfy, silvery, obtuse; leaf scars very small, half-round, slightly elevated, with 1 bundle trace; leaves lanceolate to oblong-lanceolate, entire, subacute to acute at the apex, cuneate to rounded at the base, light green above, silvery-stellate beneath, to 7.5 cm long, on silvery petioles to 1 cm long; flowers perfect, 1–3 per leaf axil, fragrant; calyx tubular, 4-lobed, the lobes about as long as the tube, silvery on the outside, yellow on the inside; stamens 4, free; fruit oval, to 12 mm long, yellow covered with silvery scales.

COMMON NAME: Russian Olive.

HABITAT: Disturbed soil.

RANGE: Native of Europe and Asia; not commonly escaped from cultivation.

ILLINOIS DISTRIBUTION: Infrequently collected in Illinois.

The Russian olive is an occasionally grown ornamental which infrequently escapes from cultivation. It differs from *E. umbellata* and *E. multiflora* by having only silvery scales on its twigs and leaves. The fruits of the Russian olive are sweet and edible.

The flowers appear in May and June.

2. **Elaeagnus umbellata** Thunb. Fl. Jap. 66. t. 14. 1784.
   Fig. 64.

Shrub to 4 m tall; twigs slender, with brown and silvery scales; buds brown, obtuse, the terminal elongated; leaf scars very small, half-round, slightly elevated, with 1 bundle trace; leaves elliptic to ovate-oblong, entire, obtuse to acute at the apex, cuneate to rounded at the base, silvery-stellate above when young, brown and silvery scaly beneath, to 7.5 cm long, on scurfy petioles to 1 cm long; flowers perfect, 1–3 per leaf axil, fragrant; calyx long-tubular,

4-lobed, the lobes much shorter than the tube, yellowish-white; stamens 4, free; fruit globose to oval, to 8 mm long, pink or red or scarlet covered with brown and silvery scales.

COMMON NAME: Autumn Olive.

HABITAT: Fields.

RANGE: Native of Asia; occasionally spreading from cultivation in the eastern United States.

ILLINOIS DISTRIBUTION: Scattered in southern Illinois. This species differs from *E. angustifolia* by having brown scales mixed with silver ones, and from *E. multiflora* by its very short stalks of the fruits.

The first adventive Illinois station discovered was along the shore of Lake of Egypt where it was found in 1969.

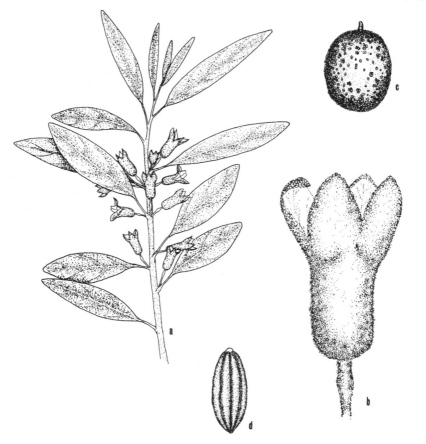

63. *Elaeagnus angustifolia* (Russian Olive). *a.* Leafy branch, with flowers, ×¾. *b.* Flower, ×5. *c.* Fruit, ×2½. *d.* Seed, ×2½.

This species is attractive to wildlife and is therefore planted frequently in Illinois.

The flowers are borne in May and June.

3. Elaeagnus multiflora Thunb. Fl. Jap. 66. 1784. *Fig. 65.*

Shrub to 4 m tall; twigs slender, with brown and silvery scales; buds brown, obtuse; leaf scars half-round, slightly elevated, with 1

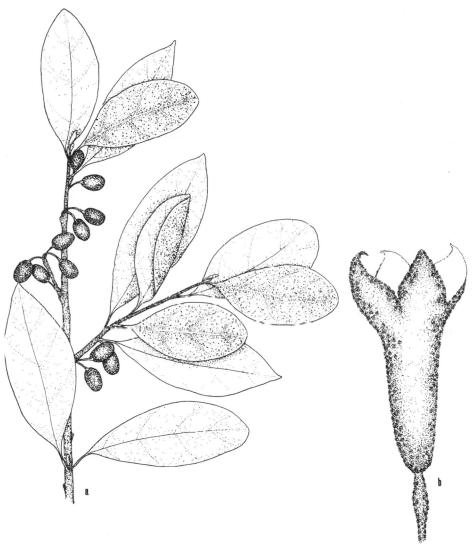

64. Elaeagnus umbellata (Autumn Olive). *a.* Leafy branch, with fruits, × ¾. *b.* Flower, × 5.

65. *Elaeagnus multiflora* (Long-stalked Oleaster). *a.* Leafy branch, with fruits, × ½. *b.* Fruit, × 2½.

bundle trace; leaves elliptic to ovate, subacute to acute at the apex, cuneate to rounded at the base, silvery-stellate above when young, silvery-stellate below and with a few brown scales, to 8 cm long, to 3 cm broad, on scurfy petioles to 1 cm long; flowers perfect, 1–2 (–4) per leaf axil, fragrant; calyx tubular, 4-lobed, the lobes about as long as the tube, silvery-brown; stamens 4, free; fruit oblong, to 15 mm long, on stalks 15–22 mm long, scarlet covered with brown and silvery scales.

COMMON NAME: Long-stalked Oleaster.

HABITAT: Disturbed soil.

RANGE: Native of China and Japan; introduced into the United States but seldom escaped.

ILLINOIS DISTRIBUTION: Known from DuPage County (Pratt's Wayne Forest Preserve, July 28, 1978, W. Lampa & L. Knowles s.n.) and Kankakee County (south of DeSelm at Kankakee River State Park, August 9, 1978, G. Wilhelm & K. Dritz 5333).

This species of Elaeagnus differs from all others in Illinois by its long-stalked fruits.

The flowers are produced in May.

Order Euphorbiales

In following the Thorne (1968) system of classification, I am including two Illinois families within this order. These are the Thymelaeaceae and the Euphorbiaceae.

In addition to these two families, Thorne assigns five additional tropical families to the Euphorbiales. These are the Pandaceae, Aextoxicaceae, Didymelaceae, Dichapetalaceae, and the Buxaceae.

Under the Cronquist (1968) system of classification, the Euphorbiaceae is grouped in the order Euphorbiales with four tropical families—Buxaceae, Daphniphyllaceae, Aextoxicaceae, and Pandaceae. The Thymelaeaceae is placed in the large order Myrtales with twelve other primarily tropical families. Included in Cronquist's Myrtales, however, are three other families represented in Illinois—Lythraceae, Onagraceae, and Melastomaceae.

THYMELAEACEAE–MEZEREUM FAMILY

Shrubs, trees, or rarely herbs; leaves usually alternate, simple, entire; flowers perfect, actinomorphic, solitary or in several types of inflorescences; calyx 4- to 5-parted, united below; petals absent (except for tropical genera); stamens usually 8 or 10, attached to the calyx; ovary superior, 1-locular, usually with 1 ovule; fruit a berry, drupe, nut, or capsule.

Approximately forty genera and nearly five hundred species comprise this family, many of which are in Australia and South Africa.

Two genera, one native and one adventive, occur in Illinois.

1. Plants woody; calyx obscurely 4-lobed; fruit a drupe _____ 1. *Dirca*
1. Plants annual herbs; calyx distinctly 4-lobed; fruit a capsule
   _____ 2. *Thymelaea*

1. *Dirca* L.–Leatherwood

Shrub with tough, fibrous bark and twigs; leaves alternate, simple, entire, deciduous, without stipules; flowers produced in clusters of

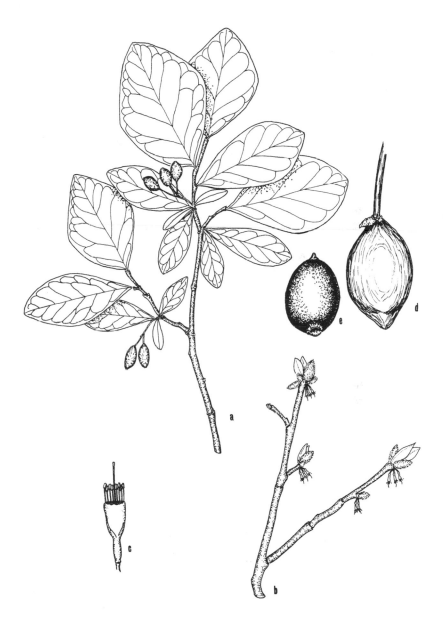

66. *Dirca palustris* (Leatherwood). *a.* Leafy branch, with fruits, ×¾. *b.* Flowering branch, ×¾. *c.* Flower, ×5. *d.* Fruit, ×5. *e.* Seed, ×5.

2–4 from twigs of the preceding season; calyx narrowly funnelform, petaloid, 4-lobed; petals absent; stamens 8, borne on the calyx; ovary superior; fruit a drupe.

One species in California and the following comprise the genus.

1. Dirca palustris L. Sp. Pl. 358. 1753. *Fig. 66*.

Shrub to 2 m tall, with fibrous bark; twigs tough and leathery; bud scales 3–4, covered with brown hairs; leaves broadly elliptic to obovate, obtuse at the apex, rounded or somewhat tapering to the base, entire, pubescent when young, becoming glabrous at maturity, up to 7 cm long; flowers borne in clusters of 2–4 on slender peduncles up to 5 mm long, produced before the leaves appear; calyx united below, inconspicuously 4-lobed above, yellow, up to 6 mm long; stamens 8, exserted above the calyx; style surpassing the stamens; drupe oval-oblong, reddish, up to 1 cm long.

COMMON NAME: Leatherwood.

HABITAT: Rich, mesic woods.

RANGE: New Brunswick to Ontario, south to Missouri and Florida.

ILLINOIS DISTRIBUTION: Scattered throughout the state, but not common.

Leatherwood is one of the more infrequently found shrubs in Illinois, where it grows primarily in rich, mesic woods. Its common name comes from the twigs which are nearly impossible to break.

This species resembles the spice-bush (*Lindera benzoin*) of the Lauraceae, but lacks the aromatic oils found in the leaves and fruits of the spice-bush.

Leatherwood flowers from late March in the southern tip of Illinois to early May in the northernmost counties.

2. Thymelaea Tourn. ex Scop.–Mezereum

Annuals (in Illinois) or perennial herbs or shrubs; leaves small, alternate; flowers bracteate, solitary or several from the axils of the leaves, sessile; sepals 4; petals absent; stamens 8, in 2 groups of 4, arising from the hypanthium; ovary superior, 1-locular; capsule 1-seeded.

Thymelaea is a genus of about 20 species, mostly in the Old World. Only the following species has been found in Illinois as an adventive.

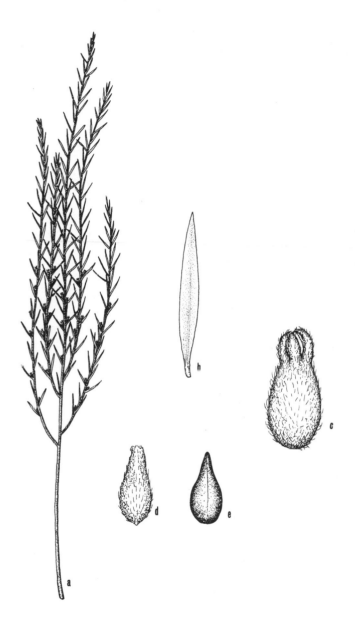

67. Thymelaea passerina (Annual Thymelaea). *a.* Habit, ×½. *b.* Leaf, ×5. *c.* Flower, ×17½. *d.* Calyx, ×10. *e.* Fruit, ×10.

1. **Thymelaea passerina** (L.) Cosson & Germain, Synop. Analit. 360. 1859. *Fig. 67.*

Stellera passerina L. Sp. Pl. 559. 1753.

Annual from fibrous roots; stems erect, slender, yellow-green, unbranched; leaves alternate, narrowly lanceolate to linear, acute at the tip, tapering to the nearly sessile base, entire, more or less leathery, glandular-punctate; flowers small, solitary or in clusters in the axils of the upper leaves, subtended by bracts; sepals 4, ovate, yellowish, much shorter than the hypanthium; petals absent; stamens 8; ovary elongated, style short, stigma capitate; fruit pear-shaped, 3 mm long; seeds brownish-black to black, ovoid.

COMMON NAME: Annual Thymelaea.
HABITAT: Disturbed soil and along a railroad.
RANGE: Native of Europe; rarely adventive in the United States.
ILLINOIS DISTRIBUTION: First discovered in Lake County in 1973; subsequently found in Kane County. The Illinois collections are among the first to be found in the United States.
This species is an unusual member of the Thymelaeaceae in that it is an annual, rather than a woody species.

EUPHORBIACEAE–SPURGE FAMILY

Herbs, shrubs, or trees, sometimes succulent and cactuslike, often with latex; leaves usually alternate, simple, stipulate; flowers unisexual, monoecious or dioecious, variously arranged; calyx present or absent, deeply 3- to 6-parted when present; corolla often absent, the parts usually 3–6 when present; stamens 1 to several, the filaments free or united; ovary superior, mostly 3-locular, with 1–2 ovules per locule, the placentation axile; flowers often arranged in a cyathium with one pistillate flower and one staminate flower composed of 1 stamen, surrounded by an involucre; fruit a capsule, berry, or drupe.

This is a large, diverse family of about three hundred genera and over eight thousand species. They are found in most parts of the World. There are twenty-two genera native to the United States.

The complex flower structure, which often is arranged in a cyathium, is difficult for a beginning student of botany to comprehend.

Many species in the family have latex. Several are cactuslike in appearance.

There are a great number of ornamentals, many grown primarily in greenhouses.

KEY TO THE GENERA OF Euphorbiaceae IN ILLINOIS

1. Leaves palmately 5- to 11-lobed; plants 3 m or more tall __ 1. *Ricinus*
1. Leaves unlobed (if 3- to 5-lobed, then the bracts pubescent); plants less than 1 m tall _____ 2
   2. Lower cauline leaves alternate, the upper opposite (*Euphorbia corollata*, with five white petallike structures per flower, may be sought here) _____ 8. *Poinsettia*
   2. Cauline leaves either all opposite or all alternate, or sometimes whorled _____ 3
3. Leaves alternate along the stem _____ 4
3. Leaves opposite along the stem _____ 9. *Chamaesyce*
   4. Uppermost leaves subtending the inflorescence whorled; latex present _____ 7. *Euphorbia*
   4. None of the leaves at the tip of the stem whorled; latex absent _ 5
5. Leaves sessile or nearly so; plants glabrous _____ 2. *Phyllanthus*
5. Leaves petiolate; plants variously pubescent _____ 6
   6. Leaves entire _____ 7
   6. Leaves variously toothed _____ 8
7. Some of the petioles at least half as long as the blades; capsule 2- to 3-celled (rarely 1-celled by abortion) _____ 3. *Croton*
7. Petioles much less than half as long as the blades; capsule 1-celled _____ 4. *Crotonopsis*
   8. Plants twining; leaves cordate _____ 5. *Tragia*
   8. Plants erect; leaves not cordate (except in *Acalypha ostryaefolia*, a plant with a 3-celled ovary) _____ 9
9. Plants with stellate pubescence; leaves with 2 glands at base of blade; pistillate flowers not subtended by cleft bracts _____ 3. *Croton*
9. Plants pubescent, but not stellate; leaves without 2 glands at base of blade; pistillate flowers subtended by cleft bracts _____ 6. *Acalypha*

1. *Ricinus* L.–Castor Bean

Annual herb or perennial shrub or tree with watery sap; leaves alternate, simple, palmately lobed, the stipules fused into a sheath; flowers unisexual, monoecious, mostly in terminal panicles; staminate flower with a 3- to 5-lobed calyx, no corolla, and up to 1,000 stamens, the filaments usually united at the base; pistillate flower with a 3- to 5-lobed calyx, no corolla, and a superior ovary with 3 locules and 1 ovule per locule; fruit a capsule.

Only the following variable species comprises the genus.

1. Ricinus communis L. Sp. Pl. 1007. 1753. Fig. 68.

Annual herb (in Illinois); stems erect, stout, to 6 m tall, glaucous, glabrous; leaves glabrous, palmately 7- to 11-lobed, the lobes serrate, ovate-oblong to lanceolate, acuminate, up to 1 m across; flowers in erect panicles, the flowers developing from the base to the apex; capsule softly echinate, 3-locular, globose, up to 2.5 cm in diameter; seeds smooth, usually mottled, with a conspicuous spongy caruncle at one end.

COMMON NAME: Castor Bean.

HABITAT: Waste ground.

RANGE: Native of Africa and India; frequently planted in Illinois but seldom escaped into waste ground.

ILLINOIS DISTRIBUTION: Scattered throughout the state, but apparently collected only in Jackson County.

This stout garden ornamental is variable in many of its characteristics. Many of these have been given horticultural names.

The seeds of the castor bean contain castor oil, a substance used as a purgative as well as in the manufacture of paints, plastics, and soaps. The seeds are extremely poisonous.

In warm temperate and tropical regions of the World, this species sometimes grows as a shrub or tree.

Castor bean flowers from June to October.

2. Phyllanthus L.

Herbs (in Illinois), shrubs, or trees, with watery sap; leaves alternate, entire, stipulate; flowers unisexual, usually monoecious, solitary or arranged in axillary cymes; staminate flower with a 4- to 6-lobed calyx, no corolla, 2–15 free or connate stamens, and a segmented disk; pistillate flower with a 4- to 6-lobed calyx, no corolla, a superior, 3-locular ovary with two ovules per locule, a segmented disk; fruit an explosively dehiscent capsule.

Phyllanthus is a genus of about seven hundred species mostly in the Tropics. Webster (1970) has written a comprehensive treatment of *Phyllanthus* in the continental United States.

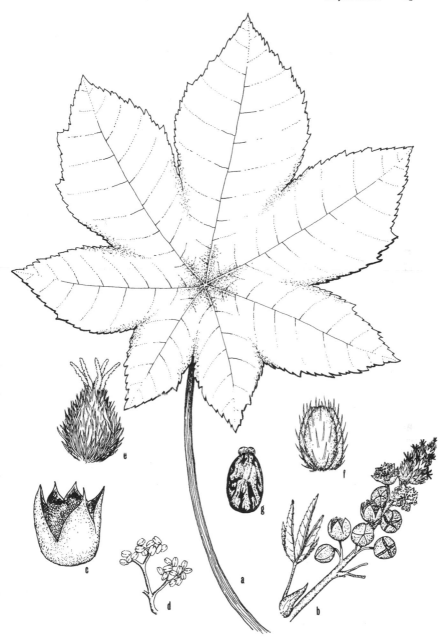

68. *Ricinus communis* (Castor Bean). *a*. Leaf, ×¼. *b*. Inflorescence, ×2½. *c*. Staminate flower, ×5. *d*. Branching stamen, ×25. *e*. Pistillate flower, ×5. *f*. Fruit, ×½. *g*. Seed, ×1.

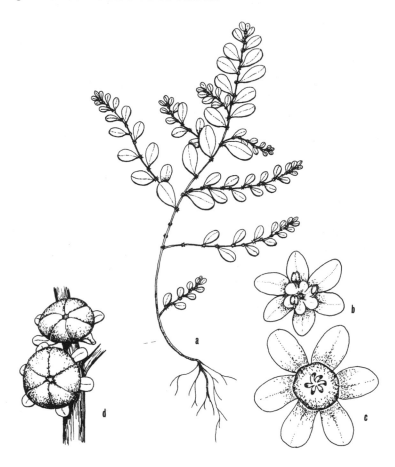

69. *Phyllanthis caroliniensis* (Phyllanthus). *a*. Habit, × 1. *b*. Staminate flower, × 10. *c*. Pistillate flower, × 10. *d*. Fruits, × 12½.

KEY TO THE SPECIES OF Phyllanthus IN ILLINOIS

1. Leaves uniform, elliptic to oblong to obovate; capsule 1.6–2.0 mm long _____ 1. *P. caroliniensis*
1. Leaves of 2 types, some of them scalelike, the others oblong; capsule 2.0–2.2 mm long _____ 2. *P. urinaria*

1. **Phyllanthus caroliniensis** Walt. Fl. Carol. 228, 1788. *Fig. 69*.
Annual or perennial herb; stems erect, simple or branched, glabrous, to 30 cm tall; leaves elliptic to oblong to obovate, obtuse and

apiculate at the apex, cuneate at the base, entire, mostly glabrous, to 20 (–30) mm long, to 10 (–15) mm broad, nearly sessile; flowers borne in axillary cymules, the cymules with one staminate and 2–3 pistillate flowers; staminate flower with a 6-lobed calyx, the lobes oblong to suborbicular, obtuse at the apex, pale yellow, 0.5–0.7 mm long, with 6 disk segments, with 3 stamens; pistillate flower with a 6-lobed calyx, the lobes linear-lanceolate, acute, green, 0.7–1.4 mm long, with an entire, cupular disk and a glabrous ovary; capsule reddish-green, 1.6–2.0 mm broad, with gray-brown seeds about 1 mm long.

COMMON NAME: Phyllanthus.
HABITAT: Mostly sandy soil.
RANGE: Pennsylvania to eastern Kansas, south to eastern Texas and Florida; West Indies.
ILLINOIS DISTRIBUTION: Occasional to rare in the southern half of the state.
Phyllanthus caroliniensis is an inconspicuous species which grows primarily in sandy, moist soils.
The flowers are produced from June to October.

2. Phyllanthus urinaria L. Sp. Pl. 982. 1753. *Fig. 70.*
Annual from tufted roots; stems prostrate to erect, usually glabrous, sometimes angular, to 50 cm long; leaves of two types, the scale leaves 2–3 mm long, spirally arranged, with denticulate, auriculate stipules, the regular leaves oblong, acute to obtuse and apiculate at the apex, more or less rounded to the asymmetrical base, hispidulous along the margins and on the veins beneath, distichously arranged, to 2.5 cm long, to 9 mm broad, with stipules to 1.5 mm long; flowers axillary, unisexual; staminate flowers in reduced cymes, subsessile, with 6 elliptic to obovate calyx lobes up to 0.5 mm long, with 6 disk segments, with 3 stamens united into a slender column; pistillate flowers solitary, subsessile, with 6 linear-oblong segments up to 1 mm long, with a 6-angled disk, with bifid styles; capsule more or less spherical, usually smooth, 2.0–2.2 mm in diameter; seeds ridged, light brown, 1.0–1.2 mm long.

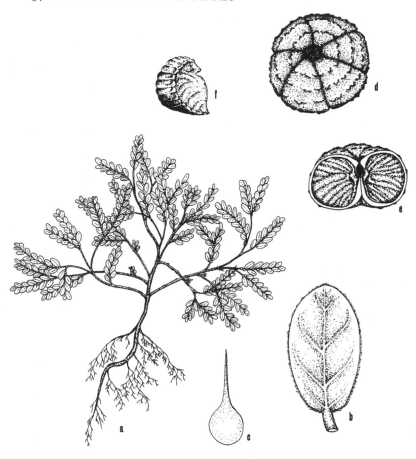

70. *Phyllanthus urinaria* (Leaf-flower). *a.* Habit, ×1. *b.* Regular leaf, ×10. *c.* Scale leaf, ×10. *d.* Fruit, ×10. *e.* Section of fruit, ×10. *f.* Seed, ×10.

COMMON NAME: Leaf-flower.

HABITAT: Disturbed soil.

RANGE: Native to the Old World tropics; adventive in the southern states, north to southern Illinois.

ILLINOIS DISTRIBUTION: Known only from Jackson County, where it was collected in Carbondale in 1977 by Mark Mohlenbrock.

This species of the Old World is commonly adventive in the southern United States. It has been found as an adventive in a nursery in Jackson County. It differs from the native *P. caroliniensis* by possessing both scale leaves and broader leaves.

Phyllanthus urinaria flowers during July and August.

3. Croton L.–Croton

Herbs (in Illinois), shrubs, or trees, without latex; leaves alternate, simple, entire, toothed, or lobed, stipulate; flowers unisexual, mostly monoecious, the pistillate usually in the lower part of the raceme, the staminate usually in the upper part; staminate flower with a 5-lobed calyx, 5 petals, an entire or segmented disk, and 3–400 free stamens; pistillate flower with a 5- to 7-lobed calyx, no corolla, an annular disk, and a 3-parted ovary; fruit a 3-parted capsule, with 1 seed per locule.

Croton is a genus of six hundred to one thousand species mostly native to North and South America. The genus is highly variable, and systematic treatments of it are just as variable.

KEY TO THE TAXA OF Croton IN ILLINOIS

1. Leaves toothed; calyx of staminate flowers 4-parted; rudimentary petals in pistillate flowers 5 _____ 1. *C. glandulosus* var. *septentrionalis*
1. Leaves entire; calyx of staminate flowers 5-parted; petals in pistillate flowers absent _____ 2
   2. Calyx of pistillate flowers 7- to 12-parted; some or all of the leaves cordate at the base _____ 2. *C. capitatus*
   2. Calyx of pistillate flowers 5-parted; leaves tapering, rounded, or barely subcordate at the base _____ 3
3. Leaves ovate to oblong; stamens 3–8; plants monoecious; seeds pitted _____ 4
3. Leaves linear to linear-oblong; stamens (8–) 10 (–12); plants dioecious; seeds reticulate _____ 5. *C. texensis*
   4. Plants silvery, with stellate-pubescence; stigmas 2-cleft; capsule 3–4 mm long, 1- to 2-celled; seeds spherical __ 3. *C. monanthogynus*
   4. Plants white, tomentose; stigmas 3-cleft; capsule 5–7 mm long, 3-celled; seeds elongated _____ 4. *C. lindheimerianus*

1. **Croton glandulosus** L. var. **septentrionalis** Muell. Arg. in DC. Prodr. 15(2):686. 1866. *Fig. 71*.

Annual from tufted roots; stems erect, branched, rough-hairy with stellate hairs and often glandular, to 75 cm tall; leaves oblong to oblong-ovate, coarsely serrate, rough-hairy, with a pair of glands at the base of the blade, to 7.5 cm long, the petiole to 2 cm long;

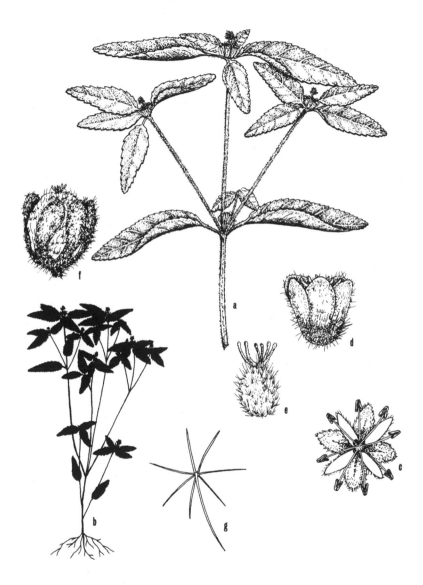

71. *Croton glandulosus* var. *septentrionalis* (Sand Croton). *a.* Leafy branch, with inflorescences, ×¾. *b.* Habit (in silhouette), ×⅛. *c.* Staminate flower, ×10. *d.* Pistillate flower, ×10. *e.* Pistil, ×12½. *f.* Fruit, ×12½. *g.* Stellate hair, ×50.

flowers in terminal or axillary spikes, the staminate above, the pistillate below; staminate flower with a 4-lobed calyx, 4 free greenish-white petals, a 4-lobed disk, and 8 free stamens; pistillate flower with 5 nearly free, rough-hairy sepals, no petals or only rudimentary petals, 3 bifid styles, and a superior ovary; capsule spherical, pubescent, up to 5 mm in diameter, containing 3 oblongoid, rugulose seeds.

COMMON NAME: Sand Croton.

HABITAT: Sandy, often disturbed, soil.

RANGE: Delaware to Kansas, south to Texas and Florida.

ILLINOIS DISTRIBUTION: Common in the southern counties, occasional in the northern counties where it is probably adventive.

I am following Fernald (1950) in considering our plants a northern variety of *C. glandulosus.* Typical var. *glandulosus* is a more southern taxon with basically larger leaves.

The flowers bloom from July to October.

2. **Croton capitatus** Michx. Fl. Bor. Am. 2:214. 1803. *Fig. 72.*

Annual herb from tufted roots; stems erect, branched, stellate- and glandular-pubescent, to 1.5 m tall; leaves lanceolate to oval, acute at the apex, rounded or subcordate at the base, entire, woolly with stellate pubescence, to 4 cm long, on stellate-pubescent petioles to 2 cm long; flowers unisexual, monoecious, in short, axillary spikes, the staminate flowers above the pistillate flowers; staminate flower with 5 nearly free, pubescent, elliptic sepals, 5 fimbriate petals which alternate with 5 glands, and (7–) 10–14 free stamens; pistillate flower with 7–12 nearly free, pubescent, elliptic sepals, no petals, and three 2- or 3-cleft styles; capsule globose, to 9 mm in diameter, with 3 flat, orbicular, gray seeds.

COMMON NAME: Capitate Croton.

HABITAT: Sandy soil; limestone glades.

RANGE: New York to Kansas, south to Texas and Georgia.

ILLINOIS DISTRIBUTION: Occasional in the southern counties, less common in the northern counties, where it is probably adventive.

Croton capitatus is distinguished from *C. glandulosus* var. *septentrionalis* by its entire leaves, from *C. mo-*

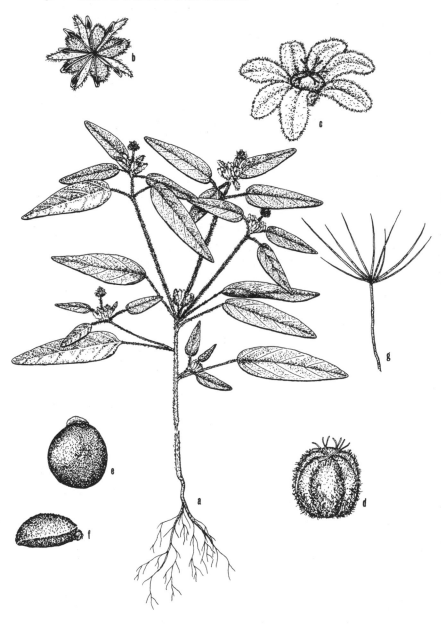

72. *Croton capitatus* (Capitate Croton). *a.* Habit, ×¾. *b.* Staminate flower,
×10. *c.* Pistillate flower, ×10. *d.* Fruit, ×2½. *e,f.* Seeds, ×5. *g.* Stellate hair,
×50.

nanthogynus by its 3-seeded fruits, from *C. lindheimerianus* by its cordate leaves, and from *C. texensis* by its monoecious condition. This species contains croton oil, a substance poisonous to cattle. The flowers are borne from July to September.

3. **Croton monanthogynus** Michx. Fl. Bor. Am. 2:215. 1803. *Fig. 73.*

Annual herb from tufted roots; stems erect, branched, silvery-stellate and glandular, to 30 cm tall; leaves oblong to ovate, obtuse at the apex, rounded or subcordate at the base, entire, silvery-stellate, to 3.5 cm long, on stellate-pubescent petioles to 1.5 cm long; flowers unisexual, monoecious, the staminate few in an erect, pedunculate cluster, the pistillate few or solitary on a recurved peduncle; staminate flower with 3–5 nearly free sepals, 3–5 free petals which alternate with 3–5 glands, and 3–8 free stamens; pistillate flower with 5 nearly free sepals, no petals, 5 glands, and 2 bifid stigmas; capsule ovoid, pubescent, 3–4 mm long, 1- to 2-locular, with 1–2 nearly round, lustrous, pitted seeds.

COMMON NAME: Croton.
HABITAT: Dry fields; bluffs.
RANGE: Virginia to Kansas, south to Texas and Georgia; Mexico.
ILLINOIS DISTRIBUTION: Occasional in the southern half of the state; apparently adventive in a few northern counties.
This species resembles the genus *Crotonopsis*, but differs by its dehiscent capsule. It grows in a variety of dry habitats.
The poisonous croton oil is produced by this species.
The flowers are borne from July to October.

4. **Croton lindheimerianus** Scheele, Linnaea 25:580. 1852. *Fig. 74.*

Perennial herb, somewhat woody near base; stems erect, branched, white-tomentose, to 45 cm tall; leaves oblong to ovate, obtuse to acute at the apex, rounded or subcordate at the base, entire, white-tomentose below, to 4.5 cm long, on tomentose petioles sometimes nearly as long as the blades; flowers unisexual, monoecious, the staminate few in an erect, pedunculate cluster, the pistillate few or solitary on a recurved peduncle; staminate flower with 3–5 nearly

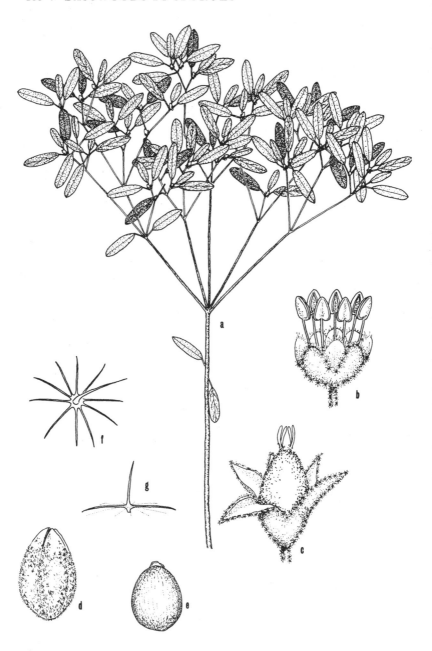

73. *Croton monanthogynus* (Croton). *a*. Habit, ×¾. *b*. Staminate flower, ×10. *c*. Pistillate flower, ×10. *d*. Capsule, ×10. *e*. Seed, ×12½. *f,g*. Stellate hairs, ×50.

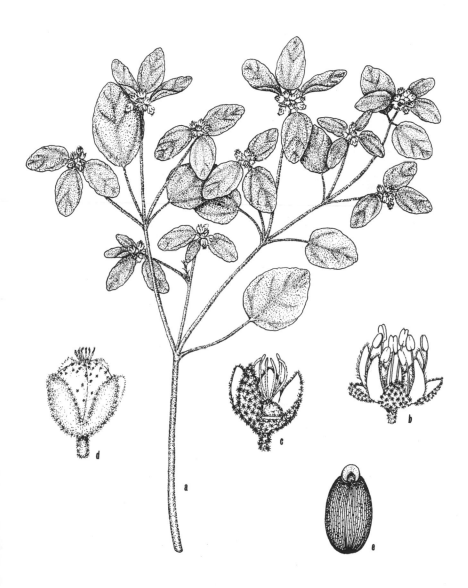

74. *Croton lindheimerianus* (Lindheimer's Croton). *a*. Upper part of plant, ×½. *b*. Staminate flower, ×17½. *c*. Pistillate flower, one sepal removed, ×17½. *d*. Fruit, ×9. *e*. Seed, ×7½.

free sepals, 3–5 free petals which alternate with 3–5 glands, and 3–8 free stamens; pistillate flower with 5 nearly free sepals, no petals, 5 glands, and 3 bifid stigmas; capsule oval, tomentose, 5–7 mm long, 3-locular, with 1 elongated, lustrous, pitted seed per locule.

COMMON NAME: Lindheimer's Croton.

HABITAT: Along a railroad (in Illinois).

RANGE: Native to the western United States, adventive in Illinois.

ILLINOIS DISTRIBUTION: Collected once as a railroad waif in Madison County.

This western species is related to the native *C. monanthogynus*, but differs by having a white rather than silvery appearance, by having a 3-locular fruit, and by having elongated seeds.

The flowers are produced from late June to October.

5. **Croton texensis** (Klotzsch) Muell. Arg. in DC. Prodr. 15(2):692. 1862. *Fig. 75*.

Hendecandra texensis Klotzsch, Erichs. Arch. 1:252. 1841.

Annual herb from tufted roots; stems erect, branched, stellate-pubescent, brownish-green, to 75 cm tall; leaves linear-oblong to lance-ovate, obtuse to acute at the apex, cuneate to rounded at the base, entire, canescent, to 7 cm long, the petioles to 4 cm long, canescent; flowers unisexual, dioecious, the staminate in axillary, pedunculate racemes, the pistillate sessile in axillary, pedunculate, capitate clusters; staminate flower with 5 nearly free sepals, no petals, 5 minute glands, and 8–12 stamens; pistillate flower with 5 nearly free sepals, no petals, 5 minute glands, and 3 bifid styles; capsule globose, tomentose, to 6 mm in diameter, the usually 3 ovoid seeds reticulate.

COMMON NAME: Texas Croton.

HABITAT: Dry soil.

RANGE: South Dakota to Wyoming, south to Arizona and Alabama; adventive in Illinois and eastward.

ILLINOIS DISTRIBUTION: Known only from Menard County (Athens, 186–, *Hall & Harbour* 514).

The above-cited specimen, which undoubtedly must represent an adventive, is in the herbarium of the Missouri Botanical Garden.

The presence of croton oil makes this a potentially dan-

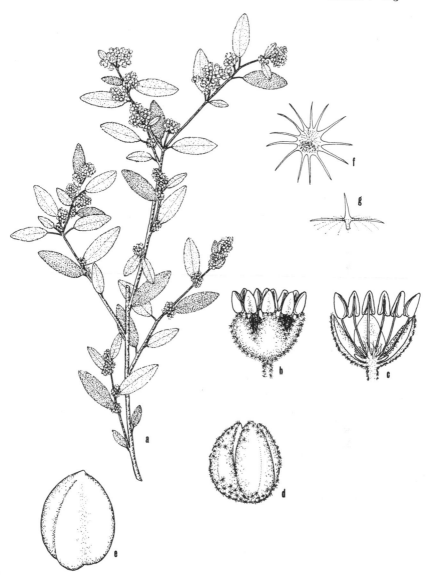

75. *Croton texensis* (Texas Croton). *a*. Habit, with staminate inflorescences, × ½. *b*. Staminate flower, × 10. *c*. Staminate flower, with the front two sepals removed, × 10. *d*. Capsule, × 5. *e*. Seed, × 10. *f,g*. Stellate hairs, × 50.

gerous species to cattle in the western United States.

The dioecious nature of this species is unique in the genus *Croton* in Illinois.

The flowers bloom from June to September.

4. *Crotonopsis* Michx.–Rushfoil

Herbs without latex; leaves alternate, simple, entire, essentially without stipules; flowers unisexual, monoecious, in racemes; staminate flower with a 5-lobed calyx, 5 free petals, and 5 free stamens; pistillate flower with a 3- to 5-lobed calyx, no petals, and a 1-locular superior ovary; fruit indehiscent, 1-locular, 1-seeded, the seed without a caruncle.

Crotonopsis is a genus of two species native to eastern North America. Species of this genus differ from *Croton monanthogynus*, which they resemble, by their indehiscent, unilocular, one-seeded fruits and their seeds without a caruncle.

KEY TO THE SPECIES OF *Crotonopsis* IN ILLINOIS

1. Fruit spinulose at tip; leaves linear-lanceolate _ _ _ _ _ _ 1. *C. linearis*
1. Fruit without spines; leaves elliptic to lanceolate _ _ _ _ 2. *C. elliptica*

1. Crotonopsis linearis Michx. Fl. Bor. Am. 2:186, pl. 46. 1803. *Fig. 76.*

Annual herb from tufted roots; stems erect, branched, to 65 cm tall, with silvery, peltate scales; leaves linear-lanceolate, acute at the apex, cuneate at the base, entire, green above, silvery-pubescent and brown-scaly below, to 3.5 cm long, on slender petioles to 5 mm long; flowers in terminal or axillary capitate clusters; staminate flower with 5 lanceolate, acute, nearly free sepals shorter than the petals; pistillate flower with 3–5 partially united, lance-ovate sepals; fruit narrowly ovoid, scaly, spinulose at the summit, to 4 mm long.

COMMON NAME: Narrow-leaved Rushfoil.

HABITAT: Sandy soil.

RANGE: South Carolina to Missouri, south to Texas and Florida; eastern Iowa; northwestern Illinois.

ILLINOIS DISTRIBUTION: Rare; known only from Carroll, Cass, LaSalle, Mason, Menard, Mercer, Scott, Tazewell, and Whiteside counties.

Pennell (1918) has presented good evidence to show that this species is specifically distinct from *C. elliptica*. The Illinois and adjacent Iowa stations for *C. linearis* are an interesting disjunction in the range of this species.

The flowers appear from July to September.

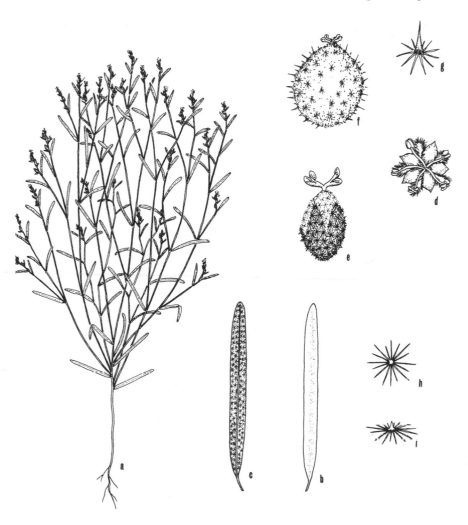

76. *Crotonopsis linearis* (Narrow-leaved Rushfoil). *a.* Habit, ×¾. *b.* Leaf, upper surface, ×2. *c.* Leaf, lower surface, ×2. *d.* Staminate flower, ×10. *e.* Pistillate flower, ×10. *f.* Fruit, ×12½. *g,h,i.* Stellate hairs, ×50.

2. Crotonopsis elliptica Willd. Sp. Pl. 4:380. 1805. *Fig. 77.*

Annual herb from tufted roots; stems erect, branched, to 40 cm tall, with silvery, peltate scales; leaves elliptic to lanceolate, acute at the apex, cuneate at the base, entire, green above with filamentous

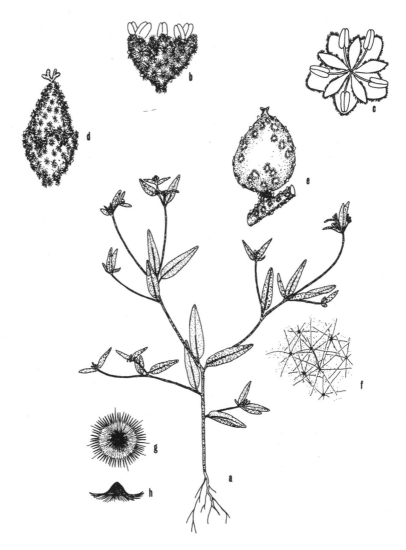

77. *Crotonopsis elliptica* (Rushfoil). *a.* Habit, ×¾. *b,c.* Staminate flowers, ×10. *d.* Pistillate flower, ×10. *e.* Fruit, ×7½. *f.* Pubescence on upper surface of leaf, ×50. *g,h.* Pubescence from lower surface of leaf, ×50.

stellate hairs, silvery-pubescent and brown-scaly below, to 2.5 cm long, on petioles to 5 mm long; flowers in terminal or axillary capitate clusters; staminate flower with 5 lanceolate, acute, nearly free sepals as long as or longer than the petals; pistillate flower with 3–

5 partially united, lance-ovate sepals; fruit narrowly ovoid, more or less scaly, not spinulose at the summit, to 3.5 mm long.

COMMON NAME: Rushfoil.
HABITAT: Dry fields, dry woods, bluffs.
RANGE: New Jersey to southeastern Kansas, south to Texas and northern Florida.
ILLINOIS DISTRIBUTION: Occasional in the southern one-third of the state.
This is one of the species typical of the dry, exposed sandstone bluff-tops of the Shawneetown Ridge across southern Illinois.
Some botanists combine this species with *C. linearis*, but I believe the differences between the two, as enumerated by Pennell (1918), are valid for specific differentiation.
The flowers appear from July to September.

5. *Tragia* L.

Perennial herbs, sometimes twining; leaves alternate, simple, entire, toothed, or lobed, stipulate; flowers unisexual, monoecious, in racemes; staminate flower with a 3- to 6-lobed calyx, no petals, and 2–5 (–50) basally connate stamens; pistillate flower with a 3- to 6-lobed calyx, no petals, and a 3-locular, hispid, superior ovary; fruit a 3-locular capsule, the seeds without a caruncle.

This genus has more than one hundred species, most of them in the tropical regions of Africa and the Americas. A revision of the United States species has been prepared by Miller and Webster (1967).

Only the following species occurs in Illinois.

1. Tragia cordata Michx. Fl. Bor. Am. 2:176. 1803. *Fig. 78.*

Tragia macrocarpa Willd. Sp. Pl. 4:323. 1805.

Perennial herb from a woody taproot; stems trailing and twining, branched, more or less hirsute, to 1.5 m or more long; leaves broadly ovate, acuminate at the apex, cordate at the base, coarsely serrate, more or less pubescent on both surfaces, ciliate, to 13 cm long, to 10 cm broad, on petioles to 8.5 cm long; stipules lance-ovate, persistent; flowers unisexual, in racemes, all but the lowest node of the raceme with staminate flowers; staminate flower with 3 (–4) pubescent calyx lobes and 3 (–4) basally connate stamens; pistillate flower with 6 (–7) pubescent calyx lobes; capsule depressed-globose, to 7 mm high, to 13 mm broad, with spherical brownish-black seeds to 5.3 mm in diameter.

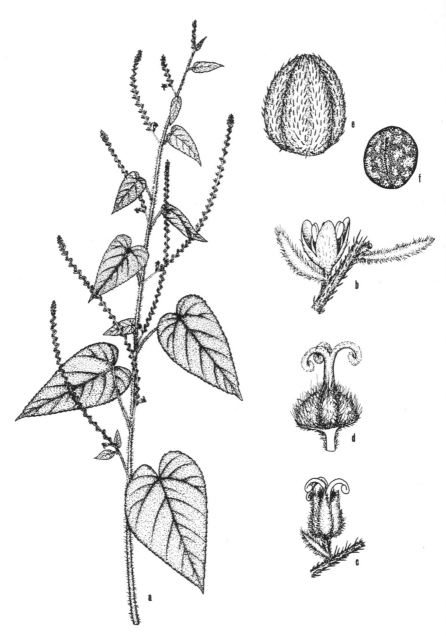

78. *Tragia cordata* (Tragia). *a*. Upper part of plant, with leaves and inflorescences, ×½. *b*. Staminate flower, ×10. *c*. Pistillate flower, ×5. *d*. Pistil, ×10. *e*. Capsule, ×5. *f*. Seed, ×7½.

COMMON NAME: Tragia.

HABITAT: Dry woods, bluffs.

RANGE: Indiana to Oklahoma, south to eastern Texas and northern Florida.

ILLINOIS DISTRIBUTION: Known only from Hardin, Johnson, and Pope counties. This species grows on dry, rocky slopes along the Ohio River near Golconda where it was first collected at this location by S. A. Forbes during the latter part of the nineteenth century.

It also occurs in swampy woods at the Heron Pond Nature Preserve in Johnson County and along limestone bluffs in Hardin County.

The flowers of this rare species bloom from July to September.

6. *Acalypha* L.–Three-seeded Mercury

Annual or perennial herbs (in Illinois), shrubs, or trees; leaves alternate, simple, toothed, stipulate; flowers unisexual, usually monoecious, in terminal or axillary spikes, the flowers subtended by usually foliaceous bracts; staminate flowers several per bract, the calyx 4-parted, petals absent, stamens 4–8, free or basally connate; pistillate flowers 1–3 per bract, the calyx 3- (5-) lobed, ovary 3-locular, superior, with 1 ovule per locule; fruit a 3-locular capsule with carunculate seeds.

Acalypha is a genus of about four hundred species, native mostly to North and South America.

KEY TO THE SPECIES OF Acalypha IN ILLINOIS

1. Leaves cordate; fruit softly echinate _____ 1. *A. ostryaefolia*
1. Leaves not cordate; fruit not softly echinate _____ 2
  2. Some or all the petioles at least ⅓ as long as the blades _____ 3
  2. Petioles at most only ¼ as long as the blades ____ 5. *A. gracilens*
3. Bracts subtending pistillate flowers 5- to 9-lobed, often glandular; stems glabrous or with incurved hairs _____ 4
3. Bracts subtending pistillate flowers 9- to 15-lobed, usually glandless; stems with spreading hairs _____ 4. *A. virginica*
  4. Capsule 3-seeded; seeds 1.2–2.0 mm long ____ 2. *A. rhomboidea*
  4. Capsule 2-seeded; seeds 2.2–3.2 mm long _____ 3. *A. deamii*

79. *Acalypha ostryaefolia* (Three-seeded Mercury). *a*. Habit, × ½. *b*. Staminate flower bud, subtended by bract, × 17½. *c*. Bract subtending pistillate flower, × 17½. *d*. Pistillate bract, opened out, × 17½. *e*. Capsule, × 5. *f*. Seed, × 10.

1. **Acalypha ostryaefolia** Riddell, Syn. Fl. W. States 33. 1835. *Fig. 79.*

Annual herb from tufted roots; stems erect, simple or branched, puberulent, to 75 cm tall; leaves ovate, acuminate at the apex, cordate at the base, serrate, to 10 cm long, on slender, glabrous or puberulent petioles to 10 cm long; staminate flowers in short, axillary spikes, the bracts subtending the flowers minute; pistillate flowers in elongated, interrupted, terminal spikes, the bracts subtending the flowers with 9–15 linear lobes; capsule depressed-globose to ovoid, 3-lobed, softly echinate, to 5 mm across, with usually 3 ovoid, wrinkled seeds about 2 mm long.

COMMON NAME: Three-seeded Mercury.

HABITAT: Riverbanks, fields, bluffs, roadsides.

RANGE: Southern Virginia to southwestern Iowa and Kansas, south to Texas and Florida; Mexico; West Indies.

ILLINOIS DISTRIBUTION: Occasional in the southern half of the state.

This species is somewhat different from the other four species of *Acalypha* in Illinois by its larger, cordate leaves, its softly echinate capsules, and its staminate and pistillate flowers which are in separate inflorescences.

Acalypha ostryaefolia is often a species of riverbanks and along streams. It may become adventive in low, cultivated fields where it can be a troublesome pest.

The flowers appear from July to October.

2. **Acalypha rhomboidea** Raf. New Fl. 1:45. 1836. *Fig. 80.*

Annual herb from tufted roots; stems erect, simple or branched, puberulent to glabrous or nearly so, to 85 cm tall; leaves lanceolate to ovate, subacute to acute at the apex, rounded or subcuneate at the base, crenate, glabrous or strigose except for the puberulent nerves on the lower surface, to 9 cm long, the glabrous or puberulent petioles at least one-third as long as the blade; staminate and pistillate flowers usually subtended by the same bract, the staminate flowers in a slender, pedunculate spike to 10 mm long, the pistillate flowers 1–3 or more at the base of the staminate peduncle; bracts deeply 5- to 9-lobed, the lobes oblong to lanceolate, ciliate, often with long-stipitate, white glands; capsule subglobose, glabrous or puberulent, 3-seeded, the seeds 1.2–2.0 mm long.

80. *Acalypha rhomboidea* (Three-seeded Mercury). *a*. Habit, ×1. *b*. Pair of inflorescences, subtended by bracts, ×7½. *c*. Bract, ×7½. *d*. Staminate calyx, in bud, ×17½. *e*. Pistillate flower, ×10. *f*. Fruit, ×10. *g*. Seed, ×25.

COMMON NAME: Three-seeded Mercury.
HABITAT: Woods, fields, bluffs, roadsides.
RANGE: Ontario to Minnesota, south to Nebraska, Texas, and Florida.
ILLINOIS DISTRIBUTION: Common; in every county. This species for years was mistakenly known as *A. virginica* in Illinois.
It is distinguished from the very similar *A. deamii* by its smaller seeds, and from *A. virginica* and *A. gracilens* by its 5- to 9-lobed bracts.

Much variation exists in the length of the petiole and the pubescence of the stems and leaves.

The flowers occur from July to October.

3. **Acalypha deamii** (Weatherby) Ahles, Vasc. Pl. Ill. 301. 1955. *Fig. 81*.

Acalypha virginica L. var. *deamii* Weatherby, Rhodora 29:197. 1927.

Acalypha rhomboidea Raf. var. *deamii* (Weatherby) Weatherby, Rhodora 39:16. 1937.

Annual herb from tufted roots; stems erect, simple or branched, puberulent or rarely villous, to 1.2 m tall; leaves broadly ovate, acute at the apex, rounded at the base, coarsely crenate, more or less pubescent, to 10.5 cm long, the usually puberulent petioles at least one-third as long as the blade; staminate and pistillate flowers usually subtended by the same bract, the staminate flowers in a slender, pedunculate spike to 12 mm long, the pistillate flowers 1– several at the base of the staminate peduncle; bracts deeply 5- to 9- lobed, the lobes oblong to lanceolate, strigose, ciliate, often with long-stipitate, white glands; capsule subglobose, usually puberulent, 2-seeded, the seeds 2.2–3.2 mm long.

COMMON NAME: Large-seeded Mercury.
HABITAT: Bottomland woods.
RANGE: Southern Indiana; eastern Illinois.
ILLINOIS DISTRIBUTION: Originally found in Vermilion Co. (along Salt Fork, three miles south of Fithian, October 19, 1952, *H. E. Ahles* 6989); also in Clark, Coles, Massac, and Montgomery counties.
I am recognizing *A. deamii* as a distinct species, following Webster (1967) in his treatment of the euphorbiaceous genera in the southeastern United States.

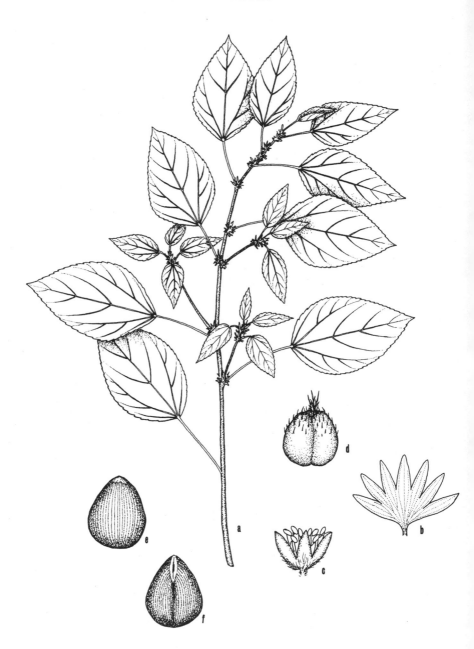

81. Acalypha deamii (Large-seeded Mercury). *a.* Upper part of plant, × ½. *b.* Bract, × 7½. *c.* Staminate flower, × 10. *d.* Fruit, × 10. *e,f.* Seed, × 25.

In addition to the larger seeds and leaves of *A. deamii*, there are only two seeds per capsule. The leaves also tend to droop.
The flowers bloom from July to October.

4. Acalypha virginica L. Sp. Pl. 1003. 1753. *Fig. 82*.

Acalypha digyneia Raf. Fl. Lud. 112. 1817.

Annual herb from tufted roots; stems erect, simple or branched, puberulent and villous, to 75 cm tall; leaves narrowly to broadly lanceolate, acute at the apex, subcuneate at the base, crenate, puberulent or nearly glabrous on both surfaces, to 8 cm long, the glabrous or puberulent petioles at least one-third as long as the blade; staminate and pistillate flowers usually subtended by the same bract, the staminate flowers in a slender, pedunculate spike to 1.7 cm long, the pistillate flowers 1–several at the base of the staminate peduncle; bracts 9- to 15-lobed, the lobes lanceolate, pubescent but usually glandless; capsule globose, 3-seeded, the seeds 1.4–1.8 mm long.

COMMON NAME: Three-seeded Mercury.
HABITAT: Fields, roadsides, bluffs, woods.
RANGE: Massachusetts to Kansas, south to Texas and Georgia.
ILLINOIS DISTRIBUTION: Occasional to common in the southern three-fourths of the state, rare and probably adventive elsewhere.
This species can usually be identified by the presence of both short, incurved hairs and spreading villi on the stems.
The bracts usually have neither white glands nor red glands. The bracts also usually have more lobes than do the bracts of other Illinois species.
The flowers bloom from July to October.

5. Acalypha gracilens Gray, Man. Bot. 408. 1848.

Annual herb from tufted roots; stems erect, simple or branched, puberulent or rarely villous, sometimes glandular-pubescent, to 80 cm tall; leaves linear to lanceolate to oblong, obtuse to acute at the apex, cuneate at the base, crenate or serrate, usually puberulent on both surfaces, to 5 cm long, the usually puberulent petioles not more than one-fourth as long as the blades; staminate and pistillate flowers usually subtended by the same bract, the staminate flowers

82. *Acalypha virginica* (Three-seeded Mercury). *a.* Habit, ×¾. *b.* Bract, ×7½. *c.* Staminate inflorescence, ×5. *d.* Staminate flower, ×17½. *e.* Pistillate flower, ×17½.

in a slender, pedunculate spike to 4 cm long, the pistillate flowers 1-several at the base of the staminate peduncle; bracts 9- to 11-lobed, the lobes oblong to deltoid, ciliate, often with either or both long-stipitate, white glands and sessile, red glands; capsule subglobose, 1- or 3-seeded, the seeds 1.3–2.0 mm long.

Two subspecies occur in Illinois.

1. Capsule 3-seeded; staminate spikes mostly 1–4 cm long _ _ _ _ _ _ _ _ _ _ _
_ 5a. *A. gracilens* ssp. *gracilens*
1. Capsule 1-seeded; staminate spikes mostly up to 1 (–1.2) cm
long _ 5b. *A. gracilens* ssp. *monococca*

5a. Acalypha gracilens Gray ssp. **gracilens** *Fig. 83a–d.*
Acalypha virginica L. var. *fraseri* Muell. Arg. Linnaea 34:44.
1865.
Acalypha virginica L. var. *gracilens* (Gray) Muell. Arg. Linnaea
34:45. 1865.
Acalypha gracilens Gray var. *fraseri* (Muell. Arg.) Weatherby,
Rhodora 29:202. 1927.
Capsule 3-seeded; staminate spikes mostly 1–4 cm long.

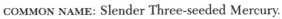

COMMON NAME: Slender Three-seeded Mercury.
HABITAT: Woods, fields, roadsides.
RANGE: New Hampshire to Wisconsin, south to Texas
and Florida.
ILLINOIS DISTRIBUTION: Occasional in the southern
half of the state, rare and possibly adventive elsewhere.
This is the common subspecies of *A. gracilens* in Illinois. The subspecies is distinguished from *A. virginica*
and *A. rhomboidea* by its shorter petioles.
I am combining var. *fraseri* with ssp. *gracilens*. Variety
fraseri supposedly differs from ssp. *gracilens* by its narrower leaves, but this difference does not seem to hold up in Illinois
material.
The time for flowering is June to September.

5b. Acalypha gracilens Raf. ssp. **monococca** (Engelm.) Web-
ster, Journ. Arn. Arb. 48:373. 1967. *Fig. 83e.*
Acalypha gracilens Raf. var. *monococca* Engelm. ex Gray, Man.
Bot., ed. 2, 390. 1856.
Capsule 1-seeded; staminate spikes mostly up to 1 (–1.2) cm long.

83. *Acalypha gracilens* (Slender Three-seeded Mercury). *a.* Habit, × ¼. *b.* Inflorescence, subtended by a bract, × 5. *c.* Staminate spike, × 1½. *d.* Seed, × 15. ssp. *monococca. e.* Staminate spike, × 1½.

COMMON NAME: Slender Three-seeded Mercury.
HABITAT: Fields, woods.
RANGE: Southern Illinois to Kansas, south to Texas and Arkansas.
ILLINOIS DISTRIBUTION: Scattered but not common in the southern two-thirds of the state.
This subspecies vegetatively looks like ssp. *gracilens*, but differs by its 1-seeded fruits and shorter staminate spikes.
The flowers are borne from June to September.

7. *Euphorbia* L.–Spurge

Herbs (in Illinois), shrubs, or trees, usually with latex; leaves alternate, opposite, or whorled, simple, often stipulate; flowers unisexual, usually monoecious; inflorescence a bisexual cyathium borne in terminal or axillary clusters; cyathium with one terminal pistillate flower and usually five staminate cymes, the staminate bracts united into an involucre; glands sometimes with petaloid appendages; staminate cymes 1- to 10-flowered, each flower composed of no perianth and one stamen; pistillate flower composed of 3–6 united sepals, or sepals absent, and a 3-locular, superior ovary; fruit a capsule.

There is considerable controversy over the species which should be attributed to *Euphorbia*. In the broad, inclusive sense of Pax and Hoffmann (1920–24), there are over 1500 species.

I prefer to follow Webster (1967) in segregating *Chamaesyce* as a distinct genus, and to follow Dressler (1962) in segregating *Poinsettia* as a distinct genus.

With species of *Chamaesyce* and *Poinsettia* removed, the remaining nine species of *Euphorbia* in Illinois fall naturally into four sections.

Section Esula. *Euphorbia cyparissias*, *E. esula*, *E. peplus*, and *E. commutata*.

Section Tithymalus. *Euphorbia helioscopia*, *E. spathulata*, and *E. obtusata*.

Section Tithymalopsis. *Euphorbia corollata*.

Section Petaloma. *Euphorbia marginata*.

KEY TO THE SPECIES OF Euphorbia IN ILLINOIS

1. Leaves entire _____ 2
1. Leaves toothed _____ 7
   2. Glands of the involucre with petallike appendages _____ 3
   2. Glands of the involucre without petallike appendages _____ 4
3. Bracts and uppermost leaves with white margins __ 1. *E. marginata*
3. Bracts and leaves green throughout _____ 2. *E. corollata*
   4. Cauline leaves linear to narrowly oblong-lanceolate, acute, sessile; seeds smooth; perennials with rhizomes _____ 5
   4. Cauline leaves obovate, obtuse or retuse, short-petiolate; seeds pitted; annuals or biennials _____ 6
5. None of the leaves over 3 mm broad _____ 3. *E. cyparissias*
5. Some or all the leaves over 3 mm broad _____ 4. *E. esula*
   6. Seeds 1.8–2.0 mm long, finely pitted throughout; capsule with rounded lobes _____ 5. *E. commutata*
   6. Seeds 1.0–1.5 mm long, with 1–4 rows of large pits; capsule with 2-keeled lobes _____ 6. *E. peplus*
7. Capsule smooth; seeds ovoid _____ 7. *E. helioscopia*
7. Capsule warty; seeds flattened or lenticular _____ 8
   8. Seeds dark brown, 1.7–2.0 mm long, obscurely reticulate; styles longer than the ovary; cyathium with 5 oblong glands _____
   _____ 8. *E. obtusata*
   8. Seeds reddish-brown, 1.3–1.5 mm long, distinctly reticulate; styles shorter than the ovary; cyathium with 4 lobes and a tuft of hairs _____ 9. *E. spathulata*

1. **Euphorbia marginata** Pursh, Fl. Am. Sept. 607. 1814. *Fig. 84.*

Dichrophyllum marginatum Klotzsch & Garcke, Monatsb. Akad. Ber. 1859:249. 1859.

Annual; stems stout, erect, branched, pilose, to 1 m tall; leaves alternate except for the whorl subtending the umbel, oblong to ovate, acute at the apex, rounded to cuneate at the base, entire, pubescent on both surfaces, to 8 cm long, short-petiolate to sessile, the uppermost leaves and the bracteal leaves bordered with white; inflorescence a terminal, 3-rayed umbel; cyathia to 4 mm long, pubescent, 5-lobed, the lobes fimbriate, the glands white-appendaged, petaloid, concave; capsule subglobose, 3-lobed, pubescent, to 7 mm in diameter, elevated above the cyathium, the seeds ovoid, dark gray, reticulate, carunculate, to 4 mm long.

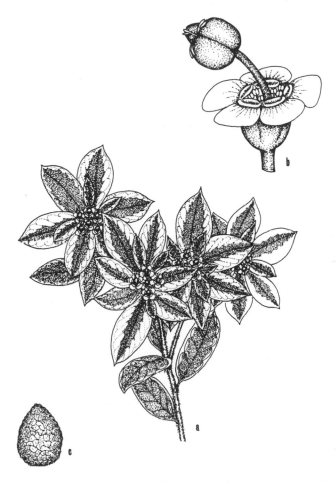

84. *Euphorbia marginata* (Snow-on-the-Mountain). *a*. Upper part of plant, × ½. *b*. Flower, × 5. *c*. Seed, × 10.

COMMON NAME: Snow-on-the-Mountain.

HABITAT: Waste areas.

RANGE: Minnesota to Montana, south to Arizona and Texas; adventive eastward.

ILLINOIS DISTRIBUTION: Occasional throughout the state.

Snow-on-the-Mountain is often cultivated as a garden ornamental because of its white-bordered bracts and upper leaves.

It is a species native to the western United States. The abundant latex produced by the plant may cause irritation when coming in contact with skin. The flowers bloom from July to October.

2. Euphorbia corollata L. Sp. Pl. 459. 1753.
Perennial herb from a stout, woody rootstock; stems 1–several from the base, erect, glabrous or softly hairy, unbranched to the inflorescence, to 1.2 m tall; leaves linear to oblong, obtuse to subacute at the apex, cuneate to the sessile or subsessile base, entire, glabrous or softly hairy, alternate except for the whorled ones subtending the inflorescence, to 5 cm long; inflorescence a several-forked umbel, panicle, or compound cyme; cyathia to 2 mm long, 5-lobed, the lobes mostly toothed, the glands white-appendaged, petaloid, rarely pinkish; capsule subglobose, 3-lobed, glabrous or pubescent, 3.5–4.5 mm in diameter, elevated above the cyathium, the seeds ovoid, minutely pitted, to 2.5 mm long.

Two varieties may be distinguished in Illinois.

1. Stem and lower surface of leaves glabrous or nearly so _____
_____ 2a. *E. corollata* var. *corollata*
1. Stem and lower surface of leaves softly hairy _____
_____ 2b. *E. corollata* var. *mollis*

2a. Euphorbia corollata L. var. **corollata** *Fig. 85a–e.*
Euphorbia paniculata Ell. Bot. S.C. & Ga. 2:660. 1824.
Euphorbia corollata L. var. *paniculata* (Ell.) Boiss. in DC. Prodr. 15(2):67. 1862.
Tithymalopsis corollata (L.) Klotzsch & Garcke, Monatsb. Akad. Ber. 1859:249. 1859.
Stem and lower surface of leaves glabrous or nearly so.

COMMON NAME: Flowering Spurge.
HABITAT: Prairies, woods, fields, roadsides.
RANGE: New York to Minnesota and Nebraska, south to Texas and Florida.
ILLINOIS DISTRIBUTION: Common; in every county.
This is the common variety of *E. corollata* in Illinois. It occurs in a great variety of habitats.
Often the petaloid glands are deformed due to a fungal defect.

Much variation occurs in regard to leaf shape, pubescence, and inflorescence size.

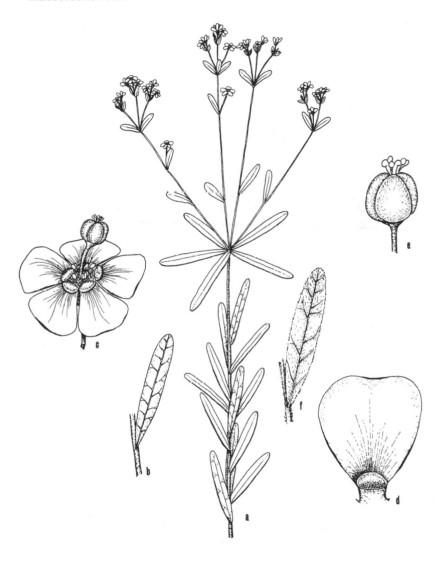

85. *Euphorbia corollata* (Flowering Spurge). *a.* Upper part of plant, with inflorescences, ×¾. *b.* Leaf, ×2. *c.* Cluster of flowers, subtended by petaloid glands, ×10. *d.* Petaloid gland, ×12½. *e.* Capsule, ×5. var. *mollis* (Hairy Flowering Spurge). *f.* Leaf, ×1¼.

I am including var. *paniculata* with var. *corollata* since the size
of the cyathia which supposedly separates the two varieties badly
overlaps, at least in Illinois specimens.
The flowers are borne from May to October.

2b. Euphorbia corollata L. var. **mollis** Millsp. Bot. Gaz.
26:267. 1898. *Fig. 85f.*

Stem and lower surface of the leaves softly hairy.

COMMON NAME: Hairy Flowering Spurge.
HABITAT: Prairies, woods, fields.
RANGE: Virginia to Oklahoma, south to Texas and Georgia.
ILLINOIS DISTRIBUTION: Occasional in the southern tip
of the state.
This southern variety is softly pubescent on most of its
vegetative parts. It sometimes occurs with the typical
variety.
The flowers are borne from May to October.

3. Euphorbia cyparissias L. Sp. Pl. 461. 1753. *Fig. 86.*

Perennial herb from tough, much-branched rhizomes; stems several together, branched or unbranched, erect to ascending, glabrous, to 75 cm tall; leaves of two kinds, those on the lower half of
the stem scalelike, those above linear, subacute at the apex, cuneate
at the sessile base, entire, glabrous, olive-green, to 3 mm broad,
alternate except for those subtending the inflorescence; umbel several-rayed, terminal and from the upper axils; cyathia to 3 mm long,
mostly 4-lobed, with 4 yellow-green, lunate glands; capsule subglobose, 3-lobed, glabrous, to 3 mm in diameter, barely elevated
above the cyathium, the seeds oblongoid, smooth, to 2 mm long,
gray-brown.

COMMON NAME: Cypress Spurge.
HABITAT: In cemeteries; along roads.
RANGE: Native of Europe; naturalized throughout most
of North America.
ILLINOIS DISTRIBUTION: Occasional in Illinois.
The extensively creeping, much-branched rhizome
permits this plant to grow in dense colonies. The olive-
green color of the leaves and stems is characteristic.
The greenish-yellow flowers are borne from May to
September.

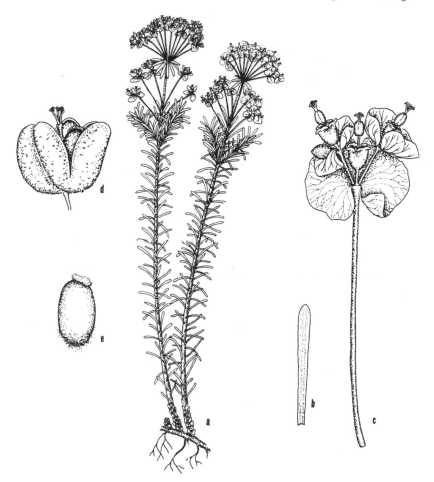

86. *Euphorbia cyparissias* (Cypress Spurge). *a.* Habit, × ½. *b.* Leaf, × 1. *c.* Inflorescences, subtended by petaloid glands, × 10. *d.* Capsule, × 7½. *e.* Seed, × 10.

4. **Euphorbia esula** L. Sp. Pl. 461. 1753. *Fig. 87.*

Tithymalus esula (L.) Hill, Hort, Kew. 174. 1768.

Euphorbia virgata Waldst. & Kit. Pl. Rar. Hung. 2:176. 1805.

Perennial herb from slender rhizomes; stems 1–several, slender, erect, glabrous, to 75 cm tall; leaves of two kinds, those on the lower half of the stem scalelike, those above linear to oblong, acute and often apiculate at the apex, cuneate to the sessile base, entire, glabrous, some or all over 3 mm broad, alternate except for those subtending the inflorescence; umbel several-rayed, terminal, with several reniform, yellow-green bracts to 1.3 cm long subtending the cyathia; cyathia to 3 mm long, mostly 4-lobed, with 4 yellowish, 2-horned glands; capsule globose, 3-lobed, glabrous, tuberculate, to 3 mm in diameter, elevated above the cyathium, the seeds oblongoid, smooth, to 2 mm long, yellow-brown.

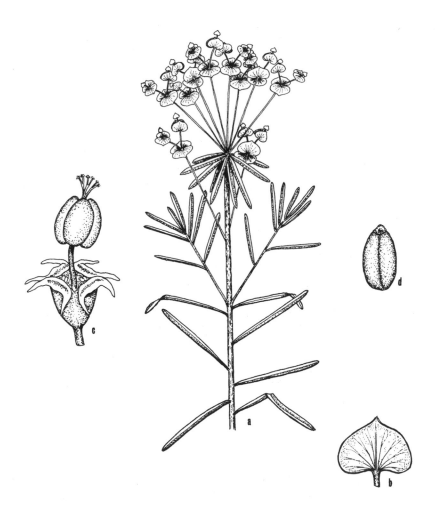

87. *Euphorbia esula* (Leafy Spurge). *a.* Upper part of plant, ×½. *b.* Bract, ×2. *c.* Flower, ×5. *d.* Seed, ×10.

COMMON NAME: Leafy Spurge.

HABITAT: Waste ground.

RANGE: Native of Europe; naturalized in much of North America.

ILLINOIS DISTRIBUTION: Occasional in the northern half of the state.

The common name of this species is derived from the leaflike bracts found among the inflorescences. This, together with the broader leaves, separates it from the closely related *E. cyparissias*.

It has the ability to run rampant in pastures.

The flowers bloom from June to September.

5. Euphorbia commutata Engelm. in Gray, Man. Bot., ed. 2, 389. 1856. Fig. 88.

Tithymalus commutatus (Engelm.) Klotzsch & Garcke, Abh. Akad. Ber. 1859:82. 1860.

Annual or biennial herb; stems decumbent at base, becoming erect, branched, slender, glabrous, to 40 cm tall; leaves obovate to spatulate, obtuse or retuse at the apex, subcuneate to rounded at the nearly sessile base, entire, glabrous, to 4 cm long, to 2 cm broad, alternate except for those subtending the inflorescence; umbel several-rayed, terminal, with the floral bracts opposite, reniform, often connate at the base; cyathia to 32.5 mm long, mostly 4-lobed, with 3–4 whitish, horned glands; capsule globose, 3-angled, glabrous, to 3 mm in diameter, elevated above the cyathium, the seeds oblongoid, pitted, to 20 mm long.

COMMON NAME: Wood Spurge.

HABITAT: Wooded slopes, along streams, and in gravelly soils.

RANGE: Pennsylvania to Minnesota, south to Texas and Florida.

ILLINOIS DISTRIBUTION: Mostly in the northern half of the state; apparently not common.

The leaves, floral bracts, and stems are sometimes reddish-tinged.

The fruits are finely pitted throughout, while the lobes or angles of the fruit are rounded instead of two-keeled as they are in *E. peplus*.

The flowers appear in May and June.

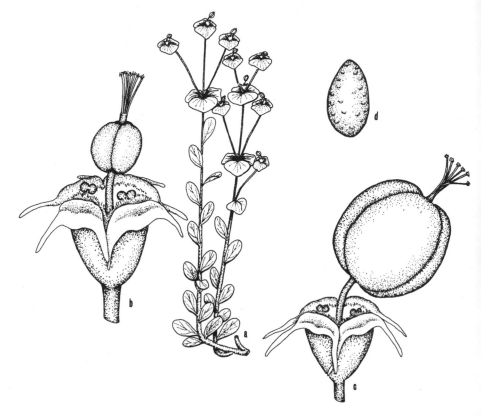

88. *Euphorbia commutata* (Wood Spurge). *a*. Habit, × ½. *b*. Flower, × 11½. *c*. Fruit, × 10. *d*. Seed, × 10.

6. Euphorbia peplus L. Sp. Pl. 456. 1753. *Fig. 89*.

Tithymalus peplus (L.) Hill, Hort. Kew. 172. 1768.

Annual herb; stems often decumbent at base, becoming erect, branched, slender, glabrous, to 30 cm tall; leaves oblong to obovate, obtuse or retuse at the apex, cuneate at the base, entire, glabrous, to 2.5 cm long, slender-petiolate, alternate except for those subtending the inflorescence; umbel several-rayed, terminal, with the floral bracts opposite, ovate; cyathia to 1.5 mm long, mostly 4-lobed, with 4 yellow horned glands; capsule globose, 3-lobed, each lobe 2-keeled, glabrous, to 2 mm in diameter, elevated above the cyathium, the seeds oblongoid, greenish-white, pitted, to 1.5 mm long.

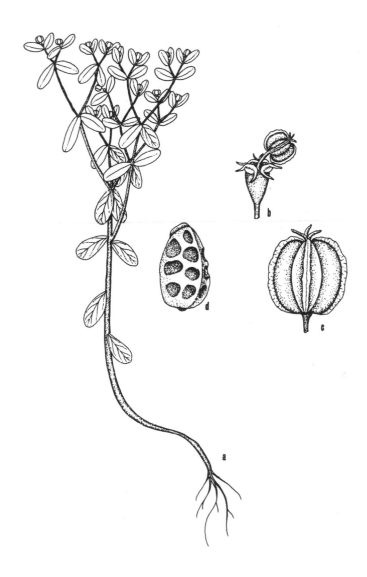

89. *Euphorbia peplus* (Petty Spurge). *a.* Habit, ×½. *b.* Flower, ×5. *c.* Fruit, ×10. *d.* Seed, ×15.

COMMON NAME: Petty Spurge.
HABITAT: Waste ground.
RANGE: Native of Europe; sparingly naturalized in eastern North America.
ILLINOIS DISTRIBUTION: Known only from a single collection made in Athens, Menard County, by Elihu Hall.
The relationship of this species clearly is with *E. commutata*, from which it differs by its smaller seeds which are pitted in 1–4 rows and by its fruits which are two-keeled on each lobe.

The petty spurge flowers from June to October. It is occasionally grown as a garden ornamental.

7. Euphorbia helioscopia L. Sp. Pl. 459. 1753. *Fig. 90.*

Tithymalus helioscopia (L.) Hill, Hort. Kew. 172. 1768.

Annual herb; stems ascending to erect, branched from near the base, glabrous or nearly so, to 45 cm tall; leaves spatulate to obovate, obtuse to retuse at the apex, cuneate to the short-petiolate base, serrulate, glabrous or nearly so, to 9 cm long, alternate except for those subtending the inflorescence; umbel several-rayed, terminal, with the floral bracts opposite, oblong to ovate, serrulate, asymmetrical at the sessile base; cyathia to 2.5 mm long, 4-lobed, with 4 yellow, oblong, stipitate glands; capsule globose to ovoid, 3-lobed, glabrous, to 4 mm in diameter, elevated above the cyathium, the seeds ovoid, reddish-brown, reticulate, to 3 mm long.

COMMON NAME: Wart Spurge; Sun Spurge.
HABITAT: Waste areas.
RANGE: Native of Europe; naturalized in eastern North America.
ILLINOIS DISTRIBUTION: Known only from Cass and Hancock counties.
This is one of three species of *Euphorbia* in Illinois with serrulate leaves. It differs from the other two, *E. obtusata* and *E. spathulata*, two native species, by its smooth fruit and ovoid seeds.

The wart spurge, sometimes known as the sun spurge, is sometimes planted as a garden ornamental.

The flowers appear from June to October.

90. *Euphorbia helioscopia* (Wart Spurge). *a*. Upper part of plant, ×½. *b*. Flower, ×10. *c*. Fruit, ×5. *d*. Seed, ×10.

8. Euphorbia obtusata Pursh, Fl. Am. Sept. 606. 1814. *Fig. 91*.

Tithymalus obtusatus (Pursh) Klotzsch & Garcke, Abh. Akad. Ber. 1859:69. 1860.

Annual herb; stem usually solitary, erect, unbranched, glabrous, yellow-green, to 90 cm tall; leaves spatulate to oblong, obtuse to subacute at the apex, cuneate to the sessile base, serrulate, glabrous, to 3 cm long, alternate except for those subtending the inflorescence; umbel several-rayed, terminal, the floral bracts opposite, ovate, serrulate, sessile; cyathia to 1 mm long, 4-lobed, with 5 oblong, red or yellow glands; styles longer than the ovary; capsule subglobose, 3-lobed, tuberculate, to 3 mm in diameter, elevated above the cyathium, the seeds lenticular, oblong, dark brown, obscurely reticulate, to 2 mm long.

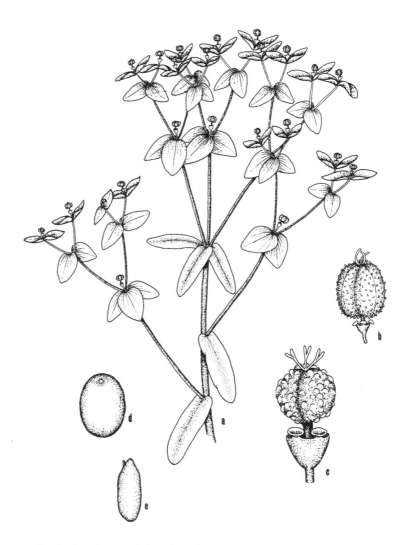

91. *Euphorbia obtusata* (Blunt-leaved Spurge). *a.* Upper part of plant, × 1. *b,c.* Flowers, × 12½. *d,e.* Seeds, × 20.

COMMON NAME: Blunt-leaved Spurge.
HABITAT: Rich woods, wooded slopes.
RANGE: Pennsylvania to Nebraska, south to eastern Texas and South Carolina.
ILLINOIS DISTRIBUTION: Occasional throughout the state, but usually not common.
This primarily southern species closely resembles *E. spathulata* from which it differs by its larger, obscurely reticulate seeds and its cyathial glands, none of which is reduced to a tuft of hairs.
The color of the glands, given by Fernald (1950) as a distinguishing character for separating *E. obtusata* from *E. dictyosperma* (= *E. spathulata*), does not seem to hold up in all Illinois specimens.
The flowers open from May to July.

9. **Euphorbia spathulata** Lam. Encycl. 2:428. 1786. *Fig.* 92.

Euphorbia dictyosperma Engelm. in Torr. Bot. Mex. Bound. Surv. 191. 1859.

Annual herb; stem usually solitary, erect, unbranched, glabrous, yellow-green, to 45 cm tall; leaves spatulate to oblong, obtuse to subacute at the apex, cuneate to the sessile base, serrulate, glabrous, to 3 cm long, alternate except for those subtending the inflorescence; umbel several-rayed, terminal, the floral bracts opposite, ovate, serrulate, sessile; cyathia to 1.5 mm long, 4-lobed, with 4 oblong, yellow or red glands and a tuft of hairs; styles shorter than the ovary; capsule subglobose, 3-lobed, tuberculate, to 3 mm in diameter, elevated above the cyathium, the seeds lenticular, oblong, reddish-brown, distinctly reticulate, to 1.5 mm long.

COMMON NAME: Spurge.
HABITAT: Limestone ledge at edge of hill prairie.
RANGE: Minnesota to Washington, south to Texas and Alabama; Mexico.
ILLINOIS DISTRIBUTION: Known only from Monroe Co. (limestone ledge, south of Fults, *J. Ozment*).
This western species barely enters Illinois on a limestone ledge a few miles east of the Mississippi River. It should be expected to occur along several of the limestone ledges in western counties of the state.
It flowers from May to July.

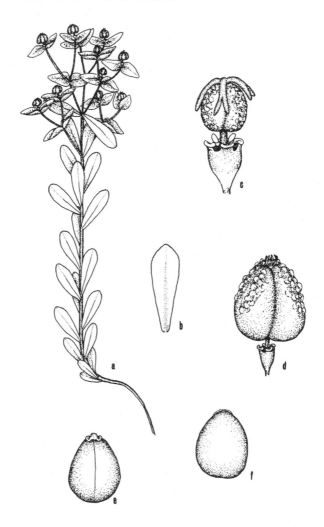

92. *Euphorbia spathulata* (Spurge). *a*. Habit, ×1. *b*. Leaf, ×2. *c,d*. Flowers, ×12½. *e,f*. Seeds, ×20.

8. *Poinsettia* Graham–Poinsettia

Herbs (in Illinois) or woody, with latex; leaves alternate or opposite, simple, with inconspicuous stipules, or stipules absent; flowers unisexual, monoecious; inflorescence a bisexual cyathium borne in terminal or axillary clusters; involucral glands mostly cuplike, reduced to usually 1, without petallike appendages; staminate flower composed of no perianth and one stamen; pistillate flower composed of 3–6 united sepals and a 3-locular, superior ovary; fruit a capsule; seeds without a caruncle.

Although *Poinsettia* is merged by many botonists with *Euphorbia*, I prefer to maintain it as a separate genus, following the reasoning of Dressler (1962). Under this concept, there are eleven species of *Poinsettia*, all New World and mostly in the tropics.

KEY TO THE SPECIES OF Poinsettia IN ILLINOIS

1. Lower leaves usually alternate; floral bracts usually red at base; seeds without a caruncle _____ 1. *P. cyathophora*
1. All leaves usually opposite; floral bracts green or cream at base, sometimes with purple spots; seeds usually carunculate ____ 2. *P. dentata*

1. **Poinsettia cyathophora** (Murr.) Klotzsch & Garcke, Monats. Akad. Berlin 1859:253. 1859.

Euphorbia cyathophora Murr. Comm. Gotting. 7:81. 1786.

Annual; stems erect, usually branched, glabrous or more uncommonly pubescent, to nearly 1 m tall; leaves alternate, at least below, mostly crowded toward the ends of the branches, linear to lanceolate to oval to panduriform, entire to undulate to toothed, glabrous or rarely pubescent, to 12 cm long, the uppermost usually with a reddish base, petiolate; cyathia terminal on the branchlets, nearly sessile, with usually 5 lobes and 1 gland, the lobes often fimbriate; capsule subglobose, 3-lobed, glabrous to puberulent, to 5 mm in diameter; seeds nearly subglobose, to 3 mm in diameter, tuberculate.

Two varieties may be distinguished in Illinois.

1. Leaves oval to lanceolate to pandurate _____ _____ 1a. *P. cyathophora* var. *cyathophora*
1. Leaves linear to narrowly lanceolate _____ _____ 1b. *P. cyathophora* var. *graminifolia*

1a. **Poinsettia cyathophora** (Murr.) Klotzsch & Garcke var. **cyathophora** Fig. 93*a–g*.

Euphorbia heterophylla L. β *cyathophora* (Murr.) Boiss. in DC. Prod. 15(2):72. 1862.

Euphorbia heterophylla L. f. *cyathophora* (Murr.) Voss in Vilmorin, Blumengartn. 1:898. 1895.

Leaves oval to lanceolate to pandurate.

93. *Poinsettia cyathophora* (Wild Poinsettia). *a.* Upper part of plant, with flowers, ×½. *b,c.* Leaf variations, ×1. *d.* Pistillate flowers, subtended by cuplike
gland, ×5. *e.* Capsule, ×7½. *f,g.* Seeds, ×7½. var. *graminifolia* (Narrow-
leaved Wild Poinsettia). *h.* Upper part of plant, with flowers, ×¼. *i.* Leaf, ×¼.

COMMON NAME: Wild Poinsettia.

HABITAT: Fields and roadsides.

RANGE: Virginia to Indiana to South Dakota, south to Texas and Florida; Mexico; West Indies.

ILLINOIS DISTRIBUTION: Occasional throughout the state.

This variety has generally been known in the past as *Euphorbia heterophylla* or *Poinsettia heterophylla*, but Dressler (1962) gives convincing evidence that these binomials belong to a distinct species whose natural range is from Arizona into tropical America.

Poinsettia cyathophora differs from *P. dentata*, the other Illinois species, by its red-based upper leaves, its alternate lower leaves, and its seeds which lack a caruncle. Leaf shape is extremely variable, ranging from broadly lanceolate to oval or even to panduriform.

The flowers bloom from July to October.

1b. **Poinsettia cyathophora** (Murr.) Klotzsch & Garcke var. **graminifolia** (Michx.) Mohlenbr. Guide Vasc. Fl. Ill. 301. 1975. *Fig. 93h–i.*

Euphorbia graminifolia Michx. Fl. Bor. Am. 2:210. 1803.
Euphorbia heterophylla L. var. *graminifolia* Engelm. Rep. U.S. & Mex. Bound. Surv. 2:190. 1859.
Poinsettia graminifolia (Michx.) Millsp. Field Mus. Pub. Bot. 2:304. 1909.

Leaves linear to narrowly lanceolate.

COMMON NAME: Narrow-leaved Poinsettia.

HABITAT: Moist, disturbed soil (in Illinois).

RANGE: Indiana to Minnesota, south to Texas and Florida.

ILLINOIS DISTRIBUTION: Known from Pope and Union counties.

Extremely narrow-leaved specimens of this variety, upon first glance, appear to be a distinct species. The structure of the cyathium and the fruits leave little doubt that the narrow-leaved plants are merely a leaf variant of *P. cyathophora.*

The variety flowers from July to October.

2. **Poinsettia dentata** (Michx.) Klotzsch & Garcke, Monatsb. Akad. Berlin 1859:253. 1859.

Euphorbia dentata Michx. Fl. Bor. Am. 2:211. 1803.

Annual; stems erect, usually branched, strigose to hispid, to 1 m tall; leaves opposite, or the very lowest sometimes alternate, linear to lanceolate to ovate to rhombic, usually dentate, strigose to hispid, to 10 cm long, the uppermost green or pale at base, petiolate; cyathia terminal on the branchlets, pedicellate, with usually 5 lobes and 1 (–4) glands, the lobes dentate; capsule subglobose, glabrous to puberulent, to 4.5 mm in diameter; seeds ovoid, angular, to 3 mm in diameter, tuberculate.

Two varieties may be distinguished in Illinois.

1. Leaves narrowly ovate to rhombic _____ 2a. *P. dentata* var. *dentata*
1. Leaves linear to narrowly lanceolate _____
_____ 2b. *P. dentata* var. *cuphosperma*

2a. **Poinsettia dentata** (Michx.) Klotzsch & Garcke var. **dentata**
Fig. 94a–e.

Leaves narrowly ovate to rhombic.

COMMON NAME: Wild Poinsettia.
HABITAT: Fields, roadsides, prairies.
RANGE: New York to South Dakota and Wyoming, south to Colorado, Texas, Louisiana, and Tennessee; Mexico to Guatemala.
ILLINOIS DISTRIBUTION: Occasional throughout the state.
This wild poinsettia differs from *P. cyathophora* by the absence of red-based bracts and the hairy stems. The typical broad-leaved variety is much more widespread than var. *cuphosperma*.

This plant may become aggressive in disturbed areas. In addition, it occurs in native habitats such as prairies.

The flowers bloom from June to September.

2b. **Poinsettia dentata** (Michx.) Klotzsch & Garcke var. **cuphosperma** (Engelm.) Mohlenbr. Guide Vasc. Fl. Ill. 301. 1975. *Fig. 94f.*

Euphorbia dentata Michx. γ *cuphosperma* Engelm. Rep. U.S. & Mex. Bound. Surv. 2:190. 1859.

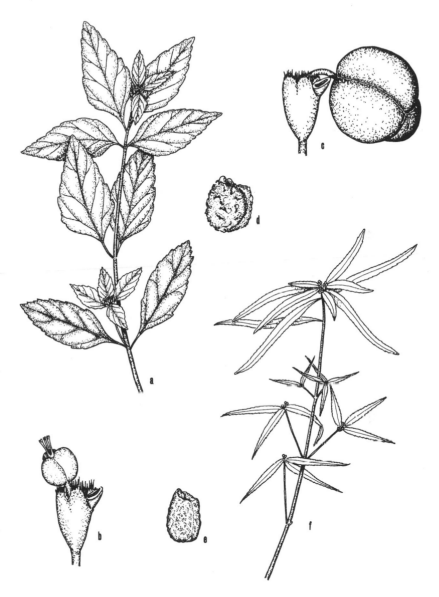

94. *Poinsettia dentata* (Wild Poinsettia). *a.* Upper part of plant, with flowers, × ½. *b.* Pistillate flower, subtended by cup-like gland, × 5. *c.* Capsule, × 7½. *d,e.* Seeds, × 10. var. *cuphosperma. f.* Upper part of plant, with flowers, × ½.

Euphorbia cuphosperma (Engelm.) Boiss. in DC. Prod. 15:73. 1862.

Poinsettia cuphosperma (Engelm.) Small, Fl. S.E.U.S. 721. 1903.

Euphorbia dentata Michx. f. *cuphosperma* (Engelm.) Fern. Rhodora 50:148. 1948.

Leaves linear to narrowly lanceolate.

COMMON NAME: Wild Poinsettia.

HABITAT: Fields and roadsides.

RANGE: Similar to var. *dentata*.

ILLINOIS DISTRIBUTION: Known only from Cook and Hancock counties.

There is some question as to the distinctiveness of var. *cuphosperma*, particularly with reference to leaf shape. Variety *cuphosperma* tends to have less tuberculate seeds than var. *dentata*.

This variety flowers from June to September.

9. Chamaesyce S. F. Gray–Spurge

Herbs (in Illinois) or shrubs, with latex; leaves opposite, simple, stipulate; flowers unisexual, monoecious; inflorescence solitary or cymose in the axils of the leaves; involucral glands 4, usually appendaged; staminate flower composed of no perianth and 1 stamen; pistillate flower composed of 3–6 united sepals and a 3-locular, superior ovary; fruit a capsule; seeds with a small caruncle.

Chamaesyce is recognized as distinct from *Euphorbia*, following the reasoning of Webster (1967), although many authors combine the two genera. *Chamaesyce* is composed of more than 200 species in temperate and tropical areas of the World.

KEY TO THE SPECIES OF Chamaesyce IN ILLINOIS

1. Ovary and capsule pubescent _____ 2
1. Ovary and capsule glabrous _____ 3
    2. Seeds with cross ridges; style 0.3–0.5 mm long, cleft less than halfway to the base _____ 1. *C. supina*
    2. Seeds smooth or minutely granular; style about 0.7 mm long, cleft halfway to the base _____ 2. *C. humistrata*
3. Leaves entire _____ 4
3. Leaves toothed _____ 6
    4. Leaves about as broad as long; capsule 1.0–1.2 mm in diameter; style

0.2 mm long; seeds about 1 mm long _____ 3. *C. serpens*
4. Leaves considerably longer than broad; capsule 2.0–3.5 mm in diameter; style 0.7–1.0 mm long; seeds 1.3–2.6 mm long _____ 5
5. Glands conspicuously appendaged; seeds 1.3–1.6 mm long; capsule 2.0–2.5 mm in diameter _____ 4. *C. geyeri*
5. Glands scarcely appendaged; seeds 2.0–2.6 mm long; capsule 3.0–3.5 mm long _____ 5. *C. polygonifolia*
6. Leaves toothed from tip to base _____ 7
6. Leaves toothed only at apex and at base _____ 8
7. Stems usually prostrate or procumbent; capsules up to 1.9 mm long; seeds with sharp angles _____ 6. *C. vermiculata*
7. Stems usually ascending to erect; capsules 1.9 mm long or longer; seeds with rounded angles _____ 7. *C. maculata*
8. Leaves broadly oblong to ovate; seeds pitted and with short cross-ridges _____ 8. *C. serpyllifolia*
8. Leaves linear-oblong; seeds with 3–6 long cross-ridges _____ _____ 9. *C. glyptosperma*

1. **Chamaesyce supina** (Raf.) Moldenke, Ann. & Class List Moldenke Coll. Numb. 135. 1939. *Fig.* 95.

Euphorbia supina Raf. Am. Monthly Mag. 2:119. 1817.

Euphorbia depressa Torr. Cat. Pl. N.Y. 45. 1819.

Annual; stems prostrate to occasionally ascending, branching from the base, puberulent to pilose to villous, spreading to 80 cm across, usually reddish; leaves opposite, elliptic to oblong to narrowly ovate, obtuse to subacute at the apex, asymmetrical at base, usually serrulate but occasionally nearly entire, pubescent to nearly glabrous, to 1.5 cm long, short-petiolate, with setiform stipules; cyathia in the axils of the leaves, the involucre 0.8–1.0 mm in diameter, pubescent, with 4 glands, with narrow red or white appendages; style 0.3–0.5 mm long, cleft less than halfway to the base; capsule subglobose, pubescent, to 2 mm in diameter, 3-angled; seeds ovoid-oblongoid, 4-angled, 0.9–1.0 mm long, with minute pits and low ridges.

COMMON NAME: Milk Spurge.

HABITAT: Disturbed or cultivated habitats.

RANGE: Quebec and Ontario to North Dakota, south to Texas and Florida.

ILLINOIS DISTRIBUTION: Common throughout the state; in every county.

This common species is closely related to *C. humistrata*

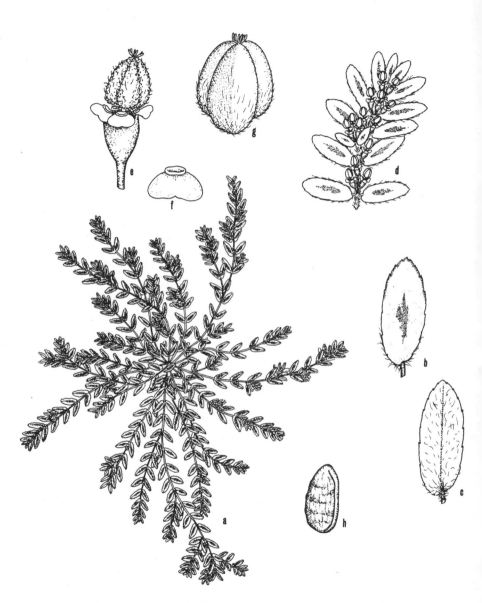

95. *Chamaesyce supina* (Milk Spurge). *a*. Habit, × ½. *b*. Leaf, showing upper surface, × 5. *c*. Leaf, showing lower surface, × 10. *d*. Leafy branch, with inflorescences, × 2½. *e*. Flowers, subtended by petaloid appendages, × 10. *f*. Petaloid appendage, × 12½. *g*. Capsule, × 12½. *h*. Seed, × 15.

on the basis of the pubescent ovaries and capsules. It differs from
C. humistrata by its ridged seeds and shorter styles.

Chamaesyce supina has long been confused with C. maculata.
Up until about the middle of the twentieth century, the binomial
C. maculata was applied to this species, while the binomial C. pres-
lii was used for what is now called C. maculata.

This species flowers from July to October.

2. **Chamaesyce humistrata** (Engelm.) Small, Fl. S.E.U.S. 713.
1903. *Fig. 96.*

Euphorbia humistrata Engelm. in Gray, Man. Bot., ed. 2, 386.
1856.

Annual; stems prostrate to occasionally ascending, branching from
the base, puberulent to pilose, spreading to 1 m across, rarely reddish; leaves opposite, ovate-oblong to elliptic-oblong, obtuse to
subacute at the apex, asymmetrical at the base, serrulate to entire,
pubescent to nearly glabrous, to 1.5 cm long, short-petiolate, with
setiform stipules; cyathia clustered in the axils of the leaves, the
involucre 0.6–0.8 mm in diameter, pubescent, with 4 glands, with
narrow red or white appendages; style about 0.7 mm long, cleft
halfway to the base; capsule subglobose, pubescent, to 2 mm in
diameter, 3-angled; seeds oblongoid, 4-angled, 0.8–1.0 mm long,
smooth or minutely granular.

COMMON NAME: Milk Spurge.

HABITAT: Sandy soil, riverbanks, disturbed soil.

RANGE: Ohio to Kansas, south to Texas and Alabama.

ILLINOIS DISTRIBUTION: Occasional throughout the
state.

Chamaesyce humistrata is closely related to C. supina
and sometimes is difficult to distinguish. It is generally
a somewhat more robust plant.

Chamaesyce humistrata occupies natural habitats in
sandy soil, primarily along rivers. Although it occasionally invades disturbed soil, it is not as aggressive as C.
supina.

This species flowers from July to September.

3. **Chamaesyce serpens** (HBK.) Small, Fl. S.E.U.S. 709. 1903.
Fig. 97.

Euphorbia serpens HBK. Nov. Gen. 2:52. 1817.

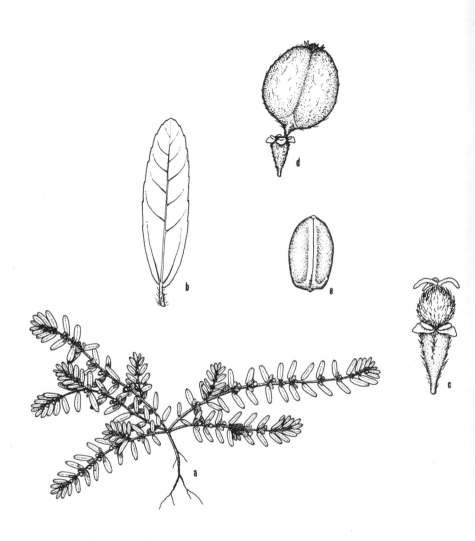

96. *Chamaesyce humistrata* (Milk Spurge). *a*. Habit, ×½. *b*. Leaf, ×5. *c*. Flower, ×15. *d*. Fruit, ×10. *e*. Seed, ×20.

Annual; stems prostrate, branching from the base, glabrous, up to 30 cm long, green or slightly glaucous; leaves opposite, ovate to oval to orbicular, obtuse at the apex, asymmetrical at the base, to 6 mm long, nearly as broad, glabrous, entire, short-petiolate, with stipules deltoid, cleft at the apex; cyathia in the axils of the leaves, the involucre 0.9–1.0 mm long, with 4 glands, with 4 narrow white

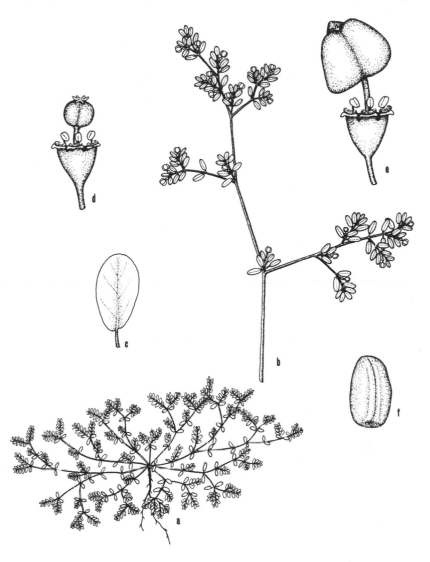

97. *Chamaesyce serpens* (Round-leaved Spurge). *a*. Habit, × ½. *b*. Upper part of plant, × 1½. *c*. Leaf, × 5. *d*. Flower, × 12½. *e*. Fruit, × 12½. *f*. Seed, × 20.

appendages; style about 0.2 mm long; capsule subglobose, glabrous, 1.0–1.2 mm in diameter; seeds ovoid, about 1 mm long, slightly 4-angled, smooth.

COMMON NAME: Round-leaved Spurge.
HABITAT: Moist, sandy soil.
RANGE: Ontario to Montana, south to New Mexico and Alabama; Mexico; West Indies.
ILLINOIS DISTRIBUTION: Scattered throughout the state, but not common.
This species has the smallest and roundest leaves of any species of *Chamaesyce* in Illinois. It is a characteristic, although not abundant, species in the moist sand along the Mississippi River south of St. Louis.
Chamaesyce serpens flowers from June to September.

4. **Chamaesyce geyeri** (Engelm. & Gray) Small, Fl. S.E.U.S. 709. 1903. *Fig. 98.*

Euphorbia geyeri Engelm. & Gray, Bost. Journ. Nat. Hist. 5:260. 1847.

Annual; stems prostrate, branching from the base, glabrous, spreading, up to 35 cm across, green; leaves opposite, elliptic to oblong, obtuse and mucronulate at the apex, asymmetrical at the base, up to 10 mm long, much longer than broad, glabrous, entire, short-petiolate, with setiform stipules; cyathia in the axils of the leaves, the involucre 0.9–1.1 mm long, with 4 glands and 4 very small white or red appendages; style 0.7–1.0 mm long; capsule subglobose, glabrous, 2.0–2.5 mm in diameter; seeds narrowly ovoid, 1.3–1.6 mm long, scarcely angular, smooth.

COMMON NAME: Geyer's Spurge.
HABITAT: Sandy soil.
RANGE: Wisconsin to North Dakota, south to Colorado, New Mexico, Texas, Illinois, and northern Indiana.
ILLINOIS DISTRIBUTION: Confined to the northwestern one-quarter of the state, and Jackson County.
The type for this species was collected by Geyer in 1842 from Beardstown, Cass County.
Chamaesyce geyeri is most abundant along the Illinois and Mississippi rivers. It is completely absent from the eastern half of the state and the southern half, except for a single collection from along the Mississippi River near Grand Tower in Jackson County.
This species most closely resembles *C. polygonifolia*, differing by its much smaller capsules and seeds.
Chamaesyce geyeri flowers from June to September.

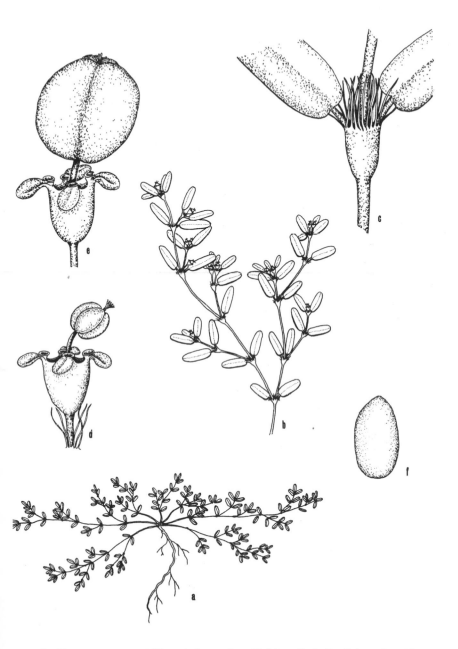

98. *Chamaesyce geyeri* (Geyer's Spurge). *a*. Habit, ×¼. *b*. Leafy branch, with inflorescences, ×¾. *c*. Node, ×5. *d*. Flower, subtended by appendages, ×10. *e*. Capsule, ×17½. *f*. Seed, ×15.

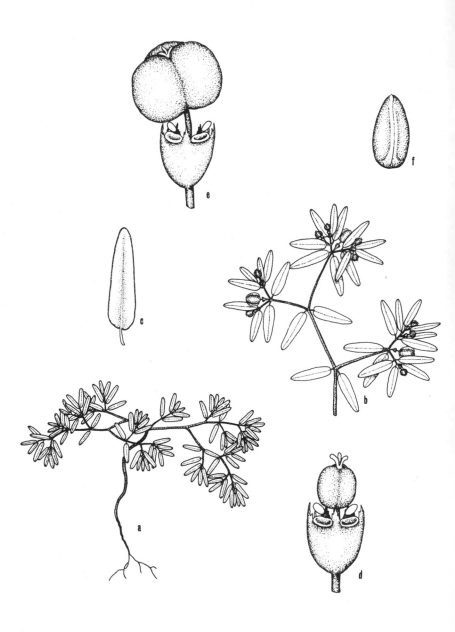

99. *Chamaesyce polygonifolia* (Seaside Spurge). *a.* Habit, × ½. *b.* Upper part of plant, × 1. *c.* Leaf, × 2½. *d.* Flower, × 8. *e.* Fruit, × 8. *f.* Seed, × 10.

5. **Chamaesyce polygonifolia** (L.) Small, Fl. S.E.U.S. 708. 1903. *Fig. 99.*

Euphorbia polygonifolia L. Sp. Pl. 455. 1753.

Annual; stems prostrate, branching from the base, glabrous, spreading, up to 20 cm across, pale green; leaves opposite, linear to oblong, obtuse to subacute and mucronulate at the apex, asymmetrical at the base, up to 15 mm long, much longer than broad, glabrous, entire, short-petiolate, with setiferous stipules; cyathia in the axils of the leaves, the involucre up to 1.5 mm long, with 4 glands, with 5 very minute appendages, or appendages absent; style 0.7–1.0 mm long; capsule subglobose, glabrous, 3.0–3.5 mm in diameter; seeds narrowly ovoid, 2.0–2.6 mm long, flattened, smooth or minutely pitted.

COMMON NAME: Seaside Spurge.

HABITAT: Dunes and beaches.

RANGE: Nova Scotia to Georgia; around the Great Lakes.

ILLINOIS DISTRIBUTION: Known only from Cook, Fulton, and Lake counties.

This species is confined to sandy beaches and dunes, and has two distinct ranges. One range is along the Atlantic Coast from Nova Scotia to Georgia. The other is around several of the Great Lakes. The Fulton County record from Illinois, although collected in sand, does not fit into either of these two ranges.

This species flowers from July to September.

6. **Chamaesyce vermiculata** (Raf.) House, Bull. N.Y. State Mus. 233–34:8. 1922. *Fig. 100.*

Euphorbia vermiculata Raf. Am. Monthly Mag. 2:119. 1817.

Euphorbia hypericifolia var. *hirsuta* Torr. Comp. Fl. N. & Mid. States 331. 1826.

Euphorbia rafinesquii Greene, Pittonia 3:207. 1897.

Euphorbia hirsuta (Torr.) Wieg. Bot. Gaz. 24:50. 1897, non Schur (1853).

Chamaesyce rafinesquii (Greene) Arthur, Torreya 11:260. 1912.

Annual; stems usually prostrate, branching from the base, hirsute or occasionally pilose, to about 30 cm long, usually green; leaves opposite, broadly lanceolate to ovate, subacute at the apex, asym-

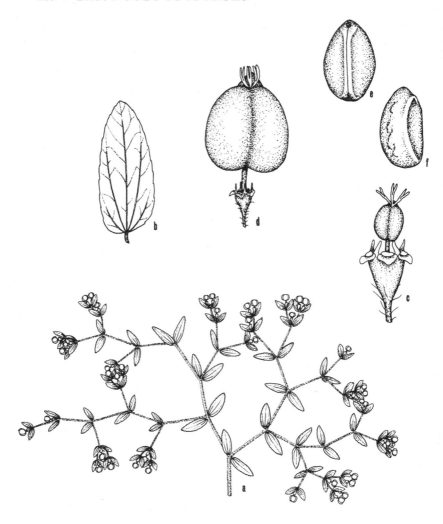

100. Chamaesyce vermicutata (Spurge). *a.* Habit, ×½. *b.* Leaf, ×2½. *c.* Flower, ×20. *d.* Fruit, ×10. *e,f.* Seeds, ×15.

metrical at the base, pilose to hirsute, up to 15 mm long, serrate all along the margin, short-petiolate, with small, deeply cleft stipules; cyathia in the axils of the leaves; involucre up to 1 mm long, with 4 glands, with white appendages; style 0.5–0.6 mm long; capsule subglobose, glabrous, 1.8–1.9 mm in diameter; seeds 1.2–1.5 mm long, sharply 4-angled, smooth or wrinkled.

COMMON NAME: Spurge.

HABITAT: Rich soil.

RANGE: Quebec to Ontario, south to northeastern Illinois, Ohio, and New Jersey; reported also from Arizona and New Mexico (Fernald, 1950).

ILLINOIS DISTRIBUTION: Known only from Lake County. The only collections of this species were made by E. J. Hill in rich soil at Lake Zurich, Lake County, on September 9, 1898.

The closest relative to this northern species is the common *C. maculata. Chamaesyce vermiculata* differs by its smaller stature and its sharply angled seeds.

This species flowers from July to September.

7. **Chamaesyce maculata** (L.) Small, Fl. S.E.U.S. 713. 1903. *Fig. 101.*

Euphorbia maculata L. Sp. Pl. 455. 1753.

Euphorbia preslii Guss. Fl. Sic. Prodr. 1:539. 1827.

Euphorbia hypericifolia Gray, Man. Bot. 407. 1848, non L. (1753).

Chamaesyce preslii (Guss.) Arthur, Torreya 2:260. 1912.

Chamaesyce lansingii Millsp. Field Mus. Publ. Bot. 2:376. 1913.

Euphorbia lansingii (Millsp.) Buhl. Bull. Chicago Acad. Sci. 5:8. 1934.

Annual; stems erect to ascending, branched, pubescent at first, often becoming glabrous, to nearly 1 m tall, green to reddish; leaves opposite, narrowly to broadly oblong, sometimes falcate, obtuse to subacute at the apex, asymmetrical at the base, glabrous to pubescent, up to 2.5 (–3.0) cm long, serrate all along the margins, often with a reddish blotch, short-petiolate, with small deltoid stipules; cyathia in the axils of the leaves; involucre 0.8–1.0 mm long, with 4 glands, with red or white appendages; style 0.7–1.0 mm long; capsule subglobose, glabrous, 1.9–2.5 mm in diameter; seeds 1.3–1.6 mm long, with rounded angles, wrinkled.

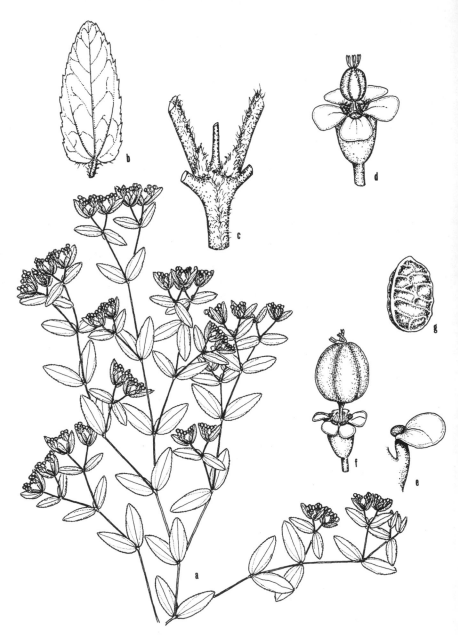

101. Chamaesyce maculata (Nodding Spurge). *a.* Habit, × ¾. *b.* Leaf, × 1½. *c.* Node, × 5. *d.* Flower, subtended by petaloid appendages, × 10. *e.* Petaloid appendage, × 10. *f.* Fruit, × 7½. *g.* Seed, × 12½.

COMMON NAME: Nodding Spurge.
HABITAT: Disturbed areas.
RANGE: Maine and Ontario to North Dakota, south to Texas and Florida; Mexico.
ILLINOIS DISTRIBUTION: Common; known from every county.
Until about 1890, most Illinois botanists called this species *Euphorbia hypericifolia* L., but Linnaeus' binomial was subsequently shown to belong to an entirely different species. Then, for about fifty years, the binomials *Euphorbia preslii* or *Chamaesyce preslii* were applied to this species. Now it has been discovered that Linnaeus' *Euphorbia maculata*, which for many years was the binomial used for *E.* or *C. supina*, is actually the species previously known as *E.* or *C. preslii*.

The reddish blotch on many of the leaves is distinctive.

This species is an aggressive weed, sometimes reaching nearly 1 meter tall. It flowers from July to September.

8. **Chamaesyce serpyllifolia** (Pers.) Small, Fl. S.E.U.S. 712. 1903. *Fig. 102*.

Euphorbia serpyllifolia Pers. Syn. 2:14. 1807.

Annual; stems prostrate to ascending, branched from the base, glabrous, to 30 cm long, green to reddish; leaves opposite, narrowly oblong to spatulate to ovate, obtuse at the apex, asymmetrical at the base, glabrous, up to 1.2 cm long, entire except for the serrulate apex and often the base, short-petiolate, with setiform stipules; cyathia from the axils of the leaves; involucre 0.8–1.0 mm long, with 4 glands, with very narrow white appendages; style 0.3–0.5 mm long; capsule subglobose, glabrous, 1.5–2.0 mm in diameter; seeds ovoid, 1.0–1.4 mm long, 4-angled, pitted and with short cross-ridges.

COMMON NAME: Spurge.
HABITAT: Sandy soil along a railroad.
RANGE: Michigan to British Columbia, south to California, New Mexico, and Texas; Mexico; adventive in Illinois.
ILLINOIS DISTRIBUTION: Collected once in sandy railroad ballast in Cook County.
This western species is adventive in Illinois. It is re-

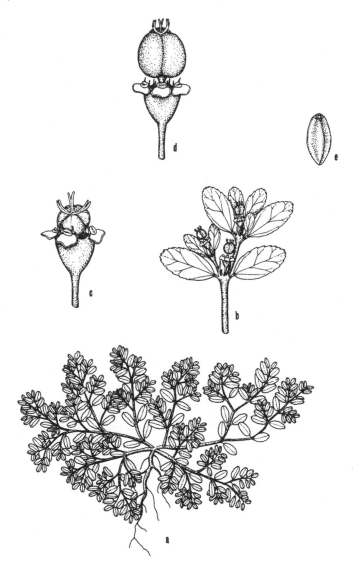

102. Chamaesyce serpyllifolia (Spurge). *a.* Habit, ×½. *b.* Flowering branch, ×3. *c.* Flower, ×10. *d.* Fruit, ×10. *e.* Seed, ×10.

lated to the native *C. glyptosperma*, differing by its broader leaves
and less ridged seeds.
Chamaesyce serpyllifolia flowers from July to October.

9. **Chamaesyce glyptosperma** (Engelm.) Small, Fl. S.E.U.S.
712. 1903. *Fig. 103.*
Euphorbia glyptosperma Engelm. Rep. U.S. & Mex. Bound.
Surv. 187. 1859.

Annual; stems prostrate to ascending, branched from near the base,
glabrous, to 30 cm long, green; leaves opposite, linear-oblong, obtuse to subacute at the apex, asymmetrical at the base, glabrous, up
to 1.2 cm long, entire except for the serrulate apex and often the
base, short-petiolate, with setiform stipules; cyathia from the axils
of the leaves; involucre 0.7–1.0 mm long, with 4 glands, with narrow white appendages; style 0.3–0.5 mm long; capsule subglobose,
glabrous, 1.4–1.8 mm in diameter; seeds oblongoid, 1.2–1.5 mm
long, 4-angled, with 3–6 long cross-ridges.

COMMON NAME: Spurge.
HABITAT: Sandy or gravelly soil.
RANGE: New Brunswick to British Columbia, south to
California, Texas, Illinois, Ohio, and New York; Mexico.
ILLINOIS DISTRIBUTION: Confined to the northern one-
third of the state and Monroe County.
This species is distinguished from all other species of
Chamaesyce in Illinois by its glabrous capsules, par-
tially toothed leaves, and heavily cross-ridged seeds.
The disjunct Monroe County specimen was collected
along the Mississippi River.
Chamaesyce glyptosperma flowers from June to September.

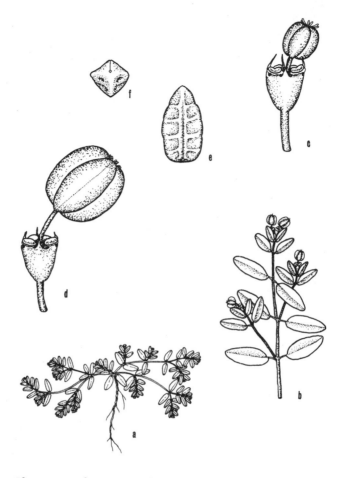

103. Chamaesyce glyptosperma (Spurge). *a.* Habit, × ½. *b.* Portion of plant, × 2. *c.* Flower, × 10. *d.* Fruit, × 10. *e,f.* Seeds, × 15.

Species Excluded

Acalypha caroliniana Walt. This binomial has been applied erroneously by Mead (1846) for *A. rhomboidea* and by Lapham (1857) and Patterson (1876) for *A. ostryaefolia*.

Anoda hastata Cav. The use of the binomial *A. hastata* by Kibbe in 1952 was undoubtedly an error for *A. cristata*.

Croton ellipticus Nutt. This species does not occur in Illinois, although Mead (1846) used this binomial for *C. monanthogynus*.

Euphorbia platyphylla L. There are several Moffatt collections of this species from Wheaton, but all are specimens collected in gardens.

Euphorbia virgata Waldst. & Kit. Hanson (1933) attributed this binomial to the leafy spurge in Illinois, instead of using the correct *E. esula*.

Hibiscus grandiflorus Michx. Patterson (1876) was the first of several nineteenth century Illinois botanists to use this binomial for *H. lasiocarpos*.

Hibiscus incanus Wendl. Palmer (1921) reported this species from southern Illinois, but he undoubtedly was referring to *H. lasiocarpos*.

Hibiscus moscheutos L. This southeastern species has been confused by several authors with our *H. palustris*.

Sphaeralcea acerifolia Torr. & Gray. The Kankakee mallow, *Iliamna remota*, was erroneously called *Sphaeralcea acerifolia* by Patterson in 1876.

Sphaeralcea rivularis (Dougl.) Torr. E. J. Hill in 1889 used the binomial of this western species incorrectly for *Iliamna remota*.

Tilia floridana Small. Palmer (1921) was the first to record this species from southern Illinois, but his specimens are *T. heterophylla*.

Tilia pubescens (Ait.) Loud. Vasey (1860, 1861) reported this species from northern Illinois, but the specimens are actually those of *T. americana* var. *neglecta*.

Ulmus campestris L. Although Snare and Hicks (1898) and Pepoon (1927) attribute this species to northern Illinois, there is no evidence that the reports are based on adventive or spontaneous plants.

Ulmus serotina Sarg. Gleason collected a specimen of an elm in sterile condition from Grand Tower in 1903 which he called *U. serotina* and which has been reported many times in the literature. The specimen is certainly not *U. serotina* and is most likely *U. alata*. A report by Sargent (1933) of a Ridgway collection of *U. serotina* from Richland County could not be verified.

Summary of the Taxa Treated in This Volume

Families	Genera	Species	Lesser Taxa
Tiliaceae	1	2	1
Sterculiaceae	1	1	
Malvaceae	11	24	1
Ulmaceae	3	10	5
Moraceae	5	7	1
Urticaceae	5	8	1
Rhamnaceae	3	9	2
Elaeagnaceae	2	4	
Thymelaeaceae	2	2	
Euphorbiaceae	9	36	4
Totals	42	103	15

GLOSSARY
LITERATURE CITED
INDEX OF PLANT NAMES

GLOSSARY

Achene. A type of one-seeded, dry, indehiscent fruit with the seed coat not attached to the mature ovary wall.

Actinomorphic. Having radial symmetry; regular, in reference to a flower.

Acuminate. Gradually tapering to a point.

Acute. Sharply tapering to a point.

Adnate. Fusion of dissimilar parts.

Alternate. Referring to the condition of structures arising singly along an axis; opposed to opposite.

Androgynous. A condition where the staminate flowers are located above the pistillate flowers.

Annual. Living for a single year.

Annular. Ringlike.

Anther. The terminal part of a stamen which bears pollen.

Apiculate. Abruptly short-pointed at the tip.

Auriculate. Bearing an earlike process.

Awned. Bearing a terminal bristle.

Axile. On the axis, referring to the place of attachment of the ovules.

Berry. A type of fruit where the seeds are surrounded only by fleshy material.

Bicarpellate. Composed of two carpels.

Biennial. Taking two years to complete a life cycle.

Bifid. Two-notched at the tip.

Bisexual. Referring to a flower which contains both stamens and pistils.

Bract. An accessory structure at the base of many flowers, usually appearing leaflike.

Caducous. Falling away early.

Calyx. The outermost ring of structures of a flower, composed of sepals.

Campanulate. Bell-shaped.

Canescent. Grayish-hairy.

Capitate. Forming a head.

Capsule. A dry, dehiscent fruit composed of more than one carpel.

Carpel. A simple pistil, or one member of a compound pistil.

Caruncle. A fleshy outgrowth near the point of attachment of a seed.

Carunculate. Bearing a caruncle.

Cauline. Belonging to a stem.

Centrifugal. Developing first at the center and then gradually toward the outside.

Ciliate. Bearing marginal hairs.

Cinereous. Ashy-gray.

Claw. A narrow, basal stalk, particularly of a petal.

Compressed. Flattened.

Concave. Curved on the inner surface; opposed to convex.

Connate. Union of like parts.

Convolute. Rolled lengthwise.

Cordate. Heart-shaped.

Coriaceous. Leathery.

Corymb. A type of inflorescence where the pedicellate flowers are arranged along an elongated axis but with the flowers all attaining around the same height.

Corymbose. Bearing a corymb.

Crenate. With round teeth.

Crenulate. With small, round teeth.

Crisped. Curled.

Crustaceous. Hard and brittle.

Cucullate. Hood-shaped.

Cuneate. Wedge-shaped or tapering to the base.

Cupular. Cup-shaped.

Cyathium. A cuplike involucre enclosing flowers.

Cyme. A type of broad and flattened inflorescence in which the central flowers bloom first.

Cymose. Bearing a cyme.

Cymule. A small cyme.

Deciduous. Falling away.

Decumbent. Lying flat, but with the tip ascending.

Dehiscent. Splitting open at maturity.

Deltoid. Triangular.

Dentate. With sharp teeth, the tips of which project outward.

Denticulate. With small, sharp teeth, the tips of which project outward.

Depressed. Flattened at the center.

Diaphragm. A partition.

Dioecious. With staminate flowers on one plant, pistillate flowers on another.

Disk. A usually fleshy, accessory part of a flower located around the ovary.

Distichous. Arranged in two vertical ranks.

Drupe. A type of fruit in which the seed is surrounded by a hard, dry covering which, in turn, is surrounded by fleshy material.

Echinate. Spiny.

Eciliate. Without marginal hairs.

Ellipsoid. Referring to a solid object which is broadest at the middle, gradually tapering to both ends.

Emarginate. Having a shallow notch at the tip.

Entire. Without teeth along the margins.

Falcate. Sickle-shaped.

Fascicle. Cluster.

Fibrous. Referring to roots borne in tufts.

Filament. That part of the stamen supporting the anther.

Filiform. Threadlike.

Fimbriate. Fringed.

Flexuous. Zigzag.

Foliaceous. Leafy.

Glabrous. Without pubescence of hairs.

Glaucous. With a whitish covering that can be rubbed off.

Globose. Round, globular.

Glomerule. A small compact cluster.

Hastate. Spear-shaped.

Hirsute. With stiff hairs.

Hispid. With short, stiff hairs.

Hispidulous. With hispid bristles.

Hypanthium. A floral cup, formed by the marginal growth of the receptacle.

Indehiscent. Not splitting open at maturity.

Inferior. Referring to the position of the ovary when it is surrounded by the adnate portion of the floral tube or is embedded in the receptacle.

Inflorescence. A cluster of flowers.

Involucel. The involucre of a secondary umbel.

Involucre. A circle of bracts which subtend a flower cluster.

Laciniate. Deeply incised.

Lanceolate. Lance-shaped; broadest near base, gradually tapering to the narrow apex.

Latex. Milky juice.

Lenticel. Corky openings on bark of twigs and branches.

Lenticular. Lens-shaped.

Lepidote. Scaly.

Ligulate. Tongue-shaped.

Linear. Elongated and uniform in width throughout.

Locular. Referring to the cavity of an ovary or a stamen.

Locule. The cavity of an ovary or an anther.

Lunate. Moon-shaped.

Membranous. Thin and transparent.

Monoecious. Bearing both sexes in separate flowers on the same plant.

Mucilaginous. Slimy.

Mucronulate. Said of a leaf with a very short, terminal point.

Obcordate. Reverse heart-shaped.

Oblanceolate. Reverse lance-shaped; broadest at apex, gradually tapering to narrow base.

Oblong. Broadest at the middle, and tapering to both ends, but broader than elliptic.

Oblongoid. Referring to a solid object which, in side view, is nearly the same width throughout, but broader than linear.

Obovate. Broadly rounded at apex, becoming narrowed below.

Obspatulate. Reverse spatulate.

Obtuse. Rounded at the apex.

Orbicular. Round.

Oval. Broadly elliptic.

Ovary. The lower swollen part of the pistil which produces the ovules.

Ovate. Broadly rounded at base, becoming narrowed above; broader than lanceolate.

Ovoid. Referring to a solid object which is broadly rounded at the base, becoming narrowed above.

Ovule. The egg-producing structure found within the ovary.

Palmate. Divided radiately, like the fingers of a hand.

Pandurate. Fiddle-shaped.

Panduriform. Fiddle-shaped.

Panicle. A type of inflorescence composed of several racemes.

Pedicel. The stalk of a flower.

Pedicellate. Bearing a pedicel.

Peduncle. The stalk of an inflorescence.

Pellucid. Clear; transparent.

Peltate. Attached away from the margin, in reference to a leaf.

Pendulous. Hanging.

Perennial. Living more than two years.

Perfect. Bearing both stamens and pistils in the same flower.

Petaloid. With color and texture of a petal.

Pilose. Bearing soft hairs.

Pinnatifid. Said of a simple leaf or leaf part which is cleft or lobed only partway to its axis.

Pistil. The ovule-producing organ of a flower normally composed of an ovary, a style, a stigma.

Pistillate. Bearing pistils but not stamens.

Pith. The soft tissue in the center of a stem.

Placentation. Referring to the manner in which the ovules are attached in the ovary.

Polygamous. With perfect and unisexual flowers on the same plant.

Procumbent. Lying on the ground.

Prostrate. Lying flat.

Puberulent. With minute hairs.

Pubescent. Bearing some kind of hairs.

Punctate. Dotted.

Punctation. A dot.

Raceme. A type of inflorescence where pedicellate flowers are arranged along an elongated axis.

Reflexed. Turned downward.

Reniform. Kidney-shaped.

Resinous. Sticky.

Reticulate. A network.

Retrorse. Pointing backward.

Retuse. Shallowly notched at a rounded apex.

Rhizome. An underground, horizontal stem.

Rhombic. Becoming quadrangular.

Rugose. Wrinkled.

Rugulose. Slightly wrinkled.

Samara. A winged fruit.

Scabrellous. Somewhat rough to the touch.

Scabrous. Rough to the touch.

Scurfy. Bearing scaly particles.

Sepal. One segment of the calyx.

Serrate. With teeth which project forward.

Serrulate. With very small teeth which project forward.

Sessile. Without a stalk.

Setaceous. Bearing bristles.

Setiform. Shaped like a bristle.

Setose. With stiff bristles.

Setulose. With short, stiff bristles.

Simple. Referring to a leaf whose blade is not divided into segments.

Spathaceous. Spathelike.

Spathe. A large sheathing bract subtending or enclosing an inflorescence.

Spatulate. Oblong, but with the basal end elongated.

Spike. A type of inflorescence where sessile flowers are arranged along an elongated axis.

Spinose. With spines.

Spinulose. Bearing small spines.

Stamen. The pollen-bearing organ of a flower, normally composed of an anther and a filament.

Staminate. Bearing stamens.

Staminodium. A sterile stamen.

Stellate. Star-shaped.

Stigma. The terminal part of a pistil which receives the pollen.

Stipitate. Bearing a stipe or stalk.

Stipulate. Bearing stipules.

Stipule. A leaflike or scaly structure found at the point of attachment of a leaf to a stem.

Stramineous. Straw-colored.

Striate. Marked with grooves.

Strigose. With appressed, straight hairs.

Style. That part of a pistil between the ovary and the stigma.

Subacute. Nearly acute.

Subcordate. Nearly heart-shaped.

Subglobose. Nearly spherical.

Subulate. With a very short, narrow point.

Superior. Referring to the position of the ovary when the free floral parts arise below the ovary.

Syncarpous. A condition where the carpels which make up a pistil are fused together.

Terete. Rounded in cross-section.

Thyrse. A crowded panicle with the lateral branches cymose.

Tomentose. Pubescent with matted wool.

Translucent. Partly transparent.

Truncate. Abruptly cut across.

Tubercle. A small, wartlike process.

Tuberculate. Bearing tubercles.

Umbel. A type of inflorescence in which the flower stalks arise from the same level.

Umbellate. Producing an umbel.

Undulate. Wavy.

Unisexual. Bearing only stamens or pistils in one flower.

Urceolate. Urn-shaped.

Valvate. Referring to floral parts which do not overlap.

Valve. One of the segments of a capsule.

Verrucose. Warty.

Villi. Long, soft, slender hairs.

Villous. With long, soft, slender, unmatted hairs.

Zygomorphic. Bilaterally symmetrical.

LITERATURE CITED

Brendel, F. 1859. The trees and shrubs of Illinois. Transactions of the State Agricultural Society 3:588–604.

Brizicky, G. K. 1964. A further note on *Ceanothus herbaceus* versus *C. ovatus*. Journal of the Arnold Arboretum 45:471–73.

Cronquist, A. 1968. The evolution and classification of flowering plants. Boston: Houghton Mifflin Co. 396 pp.

Dressler, R. L. 1962. A synopsis of Poinsettia (Euphorbiaceae). Annals of the Missouri Botanical Garden 48:329–41.

Engelmann, G. 1843. Catalogue of collection of plants made in Illinois and Missouri, by Charles A. Geyer. American Journal of Science 46:94–104.

Fernald, M. L. 1950. Gray's manual of botany. 8th ed. New York: American Book Co. 1632 pp.

Gleason, H. A. 1952. The new Britton and Brown illustrated flora of the northeastern United States and adjacent Canada. Vol. 2. New York: New York Botanical Garden. 655 pp.

Hanson, H. C. 1933. Distribution of leafy spurge in the United States. Science 78:35.

Hermann, F. J. 1946. The perennial species of *Urtica* in the United States east of the Rocky Mountains. American Midland Naturalist 35:773–78.

Jones, G. N. 1968. Taxonomy of American species of linden (*Tilia*). Illinois Biological Monographs 39:1–156.

Kibbe, A. L. 1952. A botanical study and survey of a typical midwestern county (Hancock County, Illinois). Privately published by the author at Carthage, Ill. 425 pp.

Lapham, I. A. 1857. Catalogue of the plants of the state of Illinois. Transactions of the Illinois State Agricultural Society 2:429–550.

Mead, S. B. 1846. Catalogue of plants growing spontaneously in the state of Illinois, the principal part near Augusta, Hancock County. Prairie Farmer 6:35–36, 60, 93, 119–22.

Miller, K. I., & G. L. Webster, 1967. A preliminary revision of *Tragia* (Euphorbiaceae) in the United States. Rhodora 69:241–305.

Mohlenbrock, R. H. 1975. Guide to the vascular flora of Illinois. Carbondale and Edwardsville: Southern Illinois University Press. 494 pp.

Palmer, E. J. 1921. Botanical reconnaissance of southern Illinois. Journal of the Arnold Arboretum 2:129–53.

Patterson, H. N. 1876. Catalogue of the phaenogamous and vascular cryptogamous plants of Illinois. Privately published and printed by the author at Oquawka, Ill. 54 pp.

Pax, F., & K. Hoffmann. 1920–24. Euphorbiaceae-Crotonoideae-Acalypheae-Acalyphinae, in Das Pflanzenreich 147(16):1–178.

Pennell, F. W. 1918. The genus *Crotonopsis*. Bulletin of the Torrey Botanical Club 45:477–80.

Pepoon, H. S. 1927. An annotated flora of the Chicago area. Bulletin of the Chicago Academy of Science 8:1–554.

Sargent, C. S. 1933. Manual of the Trees of North America. 3d ed. Boston and New York: Houghton Mifflin Co. 910 pp.

Shinners, L. H. 1951. *Ceanothus herbaceus* Raf. for *C. ovatus*: a correction of name. Field and Laboratory 19:33–34.

Snare, W., & E. W. Hicks. 1898. Check list of plants in the Boardman Collection, Toulon Academy. Privately published by the authors. 29 pp.

Swink, F. 1974. Plants of the Chicago region. 2d ed. Lisle, Ill.: Morton Arboretum. 474 pp.

Thorne, R. F. 1968. Synopsis of a putatively phylogenetic classification of flowering plants. Aliso 6:57–66.

Vasey, G. 1860. Additions to the Illinois flora. Prairie Farmer 22:119.

———. 1861. Additions to the flora of Illinois. Transactions of the Illinois Natural History Society 1:139–43.

Webster, G. 1967. The genera of Euphorbiaceae in the southeastern United States. Journal of the Arnold Arboretum 48:303–61.

———. 1970. A revision of *Phyllanthus* (Euphorbiaceae) in the continental United States. Brittonia 22:44–76.

INDEX OF PLANT NAMES